To my dear friend, Rochel

Mary Coleman

SEROTONIN IN DOWN'S SYNDROME

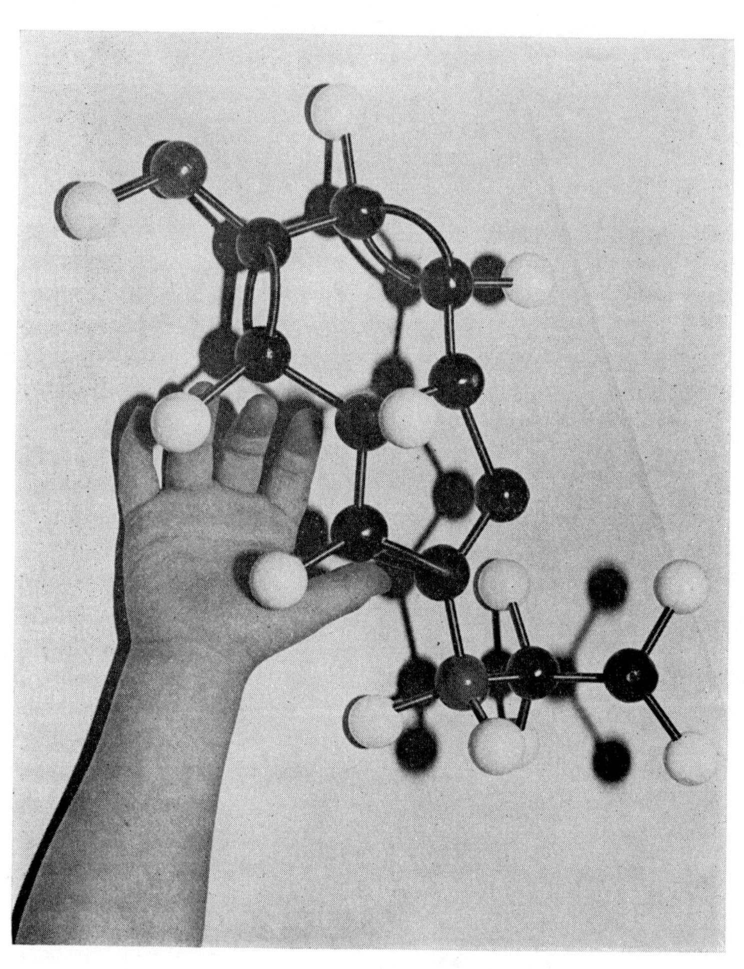

SEROTONIN IN DOWN'S SYNDROME

edited by

Mary Coleman, M.D.

1973

NORTH-HOLLAND PUBLISHING COMPANY – AMSTERDAM · LONDON

AMERICAN ELSEVIER PUBLISHING COMPANY, INC. – NEW YORK

Library of Congress Catalogue Card Number: 72–93085

ISBN North-Holland : 0 7204 4122 6

ISBN American Elsevier : 0 444 10476 3

Publishers:

North-Holland Publishing Company – Amsterdam
North-Holland Publishing Company, Ltd. – London

Sole distributors for the U.S.A. and Canada:

American Elsevier Publishing Company, Inc.
52 Vanderbilt Avenue
New York, N.Y. 10017

PRINTED IN GREAT BRITAIN

Contributors

Ann Barnet, M.D., Ann Lodge, Ph.D., Leon Cytryn, M.D.

in collaboration with

Lovisa Tatnall, M.D., Ph.D., Louis Steinberg, Ph.G., Dersh Mahanand, M.A., Betty Shanks, M.A., Paula B. Kleinfeld, A.B., Freda Hur

Illustrated by Jno Randall

dedicated to
Richmond S. Paine, M.D.

"I do not think that the relatively advanced motor development which one occasionally sees is indicative of a better than usual level of intelligence. I feel that one's estimate of the home environment, with the amount of love and stimulation which is likely to be given to the child, provides the only clue as to the possibility that a given mongol will fare somewhat better than other mongols."

R. S. Illingworth, in: *The Development of the Infant and Young Child; normal and abnormal*

Preface and Acknowledgements

This book describes a study that developed from an earlier research project of Richmond S. Paine, M.D. In collaboration with a group of investigators (Rosner, F., Ong, B. H., Mahanand, D. and Houck, J.) Paine studied a variety of intracellular enzymes and other biochemical modalities in the blood of patients with Down's syndrome. They found that an amine, also present in the brain, called serotonin (5-hydroxytryptamine, 5-HT), was depressed in the blood of this patient group. This led to the present proposal to elevate serotonin in Down's syndrome patients. The initiation and support of the work described in this research monograph was one of the final contributions made by Richmond S. Paine, brilliant and innovative pediatric neurologist, during his lifetime.

Several unusual methods (related to theoretical considerations about mental retardation) have been used in this book. Calculation of the patient's milestones during the first year of life was based on the expected date of confinement (EDC) rather than the actual date of birth (DOB). Also, a special graphing technique was used in many of the charts. The age abcissa was placed on a geometric scale, thus emphasizing the early days and weeks of the patient's life which is the period of greatest importance in creating patterns of retardation.

The studies described in this book were made possible by the devoted work and support of many people. Consistent total 5-hydroxyindole values throughout the years were made possible by the chief of the neurology laboratory, Jovita Lee, and her outstanding serotonin technologists, Delores Hijada, Felice Dorsey, Linda Greene and Judy Hamilton. Maurin Brennan and Barbara Poerkson were responsible for the catecholamine determinations and Mayada Logue for MHPG determinations. Superior chromosomal analyses were carried out in the laboratory headed by Cecil B. Jacobson, M.D. with the assistance of Dante Picciano, Richard Nelson, Lois Gladden and Kathleen Bocek. A dedicated nursing team, headed by Beatrice Rindge, R.N. was responsible for accuracy of specimen collections as well as patient counseling and support. Jacob Raitt, Ph.D. is to be commended for accurate preparation of research medicines.

Assistance to this project has been generously offered by many colleagues.

Dr. Robert Connor assisted with the statistical analyses in the manuscript and Gerhard Nellhaus, M.D. and Menek Goldstein, Ph.D. were of great assistance in their comments of the original manuscript. Office assistants, Jane Johnson and Cynthia Siemens, were invaluable. We also wish to thank David B. Coursin, M.D., David J. Boullin, Ph.D., Robert O'Brien, Ph.D., Ulli Eisner, Ph.D., Lucy Samler, M.D., Pauline Rickett, M.D., Irwin Kopin, M.D., Bernard Brodie, Ph.D. and Leonard Poryles for assistance at various stages of this research project. We also thank the following Fellows and residents: William Abernathy, David Bailey, R. E. Baska, Miryam Davis, E. S. Emery, Anne Fletcher, Jorge Holguin, Patricia Hunt, Sandra Kaplan, Donald Liegler, James Manson, James Moncrief, Yoshio Nomura, Leland Patterson, Eliott Wilner, and Gordon White.

We gratefully acknowledge the assistance of Elizabeth Bond in carrying out 3-year developmental evaluations of the infants in order to ascertain the reliability of test scores and in subsequent analyses of the data. And thanks to Dr. Lois Murphy for devising the rating scale of behavior described in Chapter 5. The technical skill of Clive Newcomb, James Meltzer, Dale Gagne, Joan Tolbert and Phyllis Evans in the EEG laboratory is to be commended.

The patients described in this monograph were referred by the following physicians:

Andrew G. Aronfy
George L. Atwell
Gordon B. Avery
A. Mynard Bacon Jr
Pat Balsamo
Ronald W. Barr
Ardwin H. Barsanti
Salvatore V. Battiata
William L. Bekenstein
Martin Berger
H. B. Berkowitz
Edwin A. Bondareff
Paul A. Bowers
Phillip C. Brunschwyler
John D. Bunce
Walter Bundy
Frederick Burke
Virginia C. Canale
George Cardany
John E. Cassidy
Robert D. Cawley
John L. Chamberlain
Joseph J. Cirotti
George J. Cohen
Allan B. Coleman
Walter A. Combs
Louis L. Cross
William Curry
Gordon W. Daisley Jr
Herbert V. Davis
Miryam M. Davis
Donald W. Dealney
Robert Detweiler
Paul A. Devore
Yvonne Driscoll
Joseph A. Dugan
Alfred Ellison
Joseph C. Evers
Robert Fidler
Donald Fishman
John J. Fitzsimons
Larry Fleischmann
Ann B. Fletcher
John L. Fox
Myron Gananian
Santiago L. Garza
Margaret J. Giannini
W. G. Gilger
Milton M. Greenberg
John W. Griffin
James E. Grimes
Vincent L. Guandolo
Richard A. Guthrie
Carl Hanfling
Roger W. Hedin
Harold Hobart
Leroy E. Hoeck
Richard J. Hollander
William H. Howard
Walter L. Jacobs
Cecil B. Jacobson
M. A. Jansa
Albert J. Kanter
John F. Kelly
Ronald E. Keyser
Robert J. Knerr
H. Donald Knox
Frank Kopack
Clarence Laing
Raymond Latham
Leonard Lefkowitz
Charles Leslie

Norman H. Liebshutz
Anthony D. Lutkus
Iradj Mahdavi
Andrew Margileth
Francis M. Mastrota
James D. McDowell
William Mebane
John Menkes
Arthur Z. Metzger
Morris I. Michael
Albert Modlin
Marvin I. Mones
Daniel L. Moore
Frank J. Murphy
John J. Murray
Gerhard Nellhaus
Karin B. Nelson
Robert Nelson
Frank W. Neuberger
Frederick North Jr
John O'Connell
Vincent L. O'Donnell
Beale H. Ong
Mark N. Ozer
Lawrence C. Pakula
Robert Parr
Frank S. Pellegrini
Dennis Penn
Robert Reid
Protacio A. Reyes
Edmond Rodriguez
Sidney Ross
Sandra K. Sabo
C. Francis Scaless
Robert T. Scanlon
Louis Schaner
John Schriner
Sidney B. Seidman
John C. Seymour
Ben Shaver
C. G. Shellenberger
Bennett Sherman
Robert L. Simpson
William H. Simpson
Margaret Snow
William J. Sohn
Herbert M. Solomon
Catherine Spears
George R. Spence
Douglas M. Spencer
J. Ward Stackpole
Richard Stein
Stanley Steinberg
Ralph Stiller
Donald Straus
Albert J. Strauss
Yas Takagi
Robert T. Thibadeau
William F. Thompson
Beno S. Vajda
Louis P. Vecchiarello
Beverly Vander Veer
Francisco Venegas
Robert Orr Warthen
Charles R. Webb
David L. Weinstein
Milton W. Werthman Jr
Charles Wiley
John C. Williams
Ben D. Wilmot
Agnes C. Wilson
Thomas A. Wilson
Robert K. Wineland
Mitchell Woldoff
Stanley I. Wolf
Harold T. Yates
Richard A. Young

The excellent cooperation of The George Washington Audio-Visual Department, Image Inc., and Robert Der made possible the photographs appearing in this book.

The funding of this project was primarily through research grants. However, the continuity and completion of the project was made possible through the dedicated and generous efforts of the Mothers of Young Mongoloids. Paula Felder, Celia Wyman, Marilyn Trainer, and Maria Hartman provided the leadership for these fund drives. Also, contributions were made by: Donald G. Agger, Association for Children with Down's Syndrome, Mrs. John Ahlers, Albert Beekhuis Foundation, Coulter Electronics, Gage Fund, Lillian F. Keefe, Landegger Foundation, Henry J. Leir, Charles E. Merrill Trust, Montgomery County (of Maryland) Association for Retarded Children, Montgomery County (of Pennsylvania) Association for Retarded Children, Northern Virginia Association for Retarded Children, Singer Freidan Corporation, and the Sperry Rand Corporation.

The amino acid used in this research study was most generously supplied by Calbiochem.

Grant support of this project has come from the Division of Mental Retardation of the Bureau of Health Services of HEW (contract #PH110-44), Maternal and Child Health grant #H256 and H256-C1, N.I.N.D.S. grant

#NB08429, Division of Research Resources grant #RR00284, N.I.C.H.H.D. grant #HDo2296, NIMH Career Development Awards #K2-45,472 and MH40796-05. Mrs. Gloria Wackernah and Dr. Robert Jaslow helped make this possible.

Contents

	Introduction *Mary Coleman*	1
Chapter 1	Baseline serotonin levels in Down's syndrome patients *Mary Coleman and Dersh Mahanand*	5
Chapter 2	Administration of 5-hydroxytryptophan—open studies *Mary Coleman*	25
Chapter 3	A double blind trial of 5-hydroxytryptophan in trisomy 21 patients *Mary Coleman and Louis Steinberg*	43
Chapter 4	Early behavioral development in Down's syndrome *Ann Lodge and Paula B. Kleinfeld*	61
Chapter 5	Personality development in patients on the double blind study receiving 5-hydroxytryptophan or placebo *Leon Cytryn and Lovisa Tatnall*	87
Chapter 6	EEG and sensory evoked potentials *Ann B. Barnet and Betty L. Shanks*	95
Chapter 7	Acute and chronic side effects of 5-hydroxytryptophan *Mary Coleman and Ann B. Barnet*	117
Chapter 8	Other methods of raising serotonin *Mary Coleman*	135
Chapter 9	Platelet serotonin in disturbances of the central nervous system *Mary Coleman and Freda Hur*	149
Chapter 10	Summary	165
	Appendices I–IX	169
	Subject index	221

Introduction

Children with Down's syndrome are a group of retarded patients more commonly known as mongols or mongoloids because of the misinterpreted 'oriental' slant of their eyes. The inhabitants of Mongolia, however, have protested about the use of the term mongol to designate retarded children, leading the medical community to seek a new name for these patients. Dr. J. Langdon Down (1866) of London originally differentiated this group of patients from cretins (or infants with hypothyroidism); this is the basis of the choice of the term 'Down's syndrome' to describe the patients.

As children, Down's syndrome patients have a group of distinguishing characteristics—round heads, upward slanting eyes, underdevelopment of the face, protruding tongue, short stature and obesity—that make them identifiable to observers. More than 100 stigmata, or structural changes, can be found in this patient group as a whole although a single individual may have as few as 10 of the stigmata of the syndrome.

French investigators (Lejeune *et al.*, 1959) described the presence of an extra chromosome in the leukocytes of these patients. It was a member of the 21st–22nd group of chromosomes and the extra chromosome in these patients has now been designated as the 21st. Chromosomes are usually paired; patients with Down's syndrome may have a set of three of the 21st chromosomes; hence their designation as 'trisomy 21'. A small percentage (10% or less) of the patients do not have an extra chromosome in every cell. One subgroup is 'mosaic' patients, who have a mixture of normal cells and cells containing the extra 21st chromosome. Another subgroup are translocation patients who have a large chromosome, thought to be a combination of most of the genetic material from a 21st chromosome with most of the material of another regular chromosome.

Studies in the first half of this century consistently reported that these patients were severely retarded. Even as late as 1954, a paper entitled 'Two mongols of usually high mental status' described patients with Stanford-Binet Intelligence Quotient Scores of 42 and 44 (Strickland, 1954). Since then, more and more patients with moderate, mild or even normal intelligence have been reported. There have been many mosaics recorded with normal intelligence; also normal I.Q. levels are described in three translocation patients (Finley *et al.*, 1965; Carter, 1967; Shaw, 1962). Seventy-four (Wechsler Adult Intelligence Scale) is the highest intelligence quotient reported in a patient with trisomy 21 that has remained at a higher level into adult life; this patient described by Zellweger *et al.* (1968) was 26 years old when tested. Another superior trisomy 21 patient is Nigel Hunt, author of a book (Hunt, 1967), a boy with outstanding verbal ability who scored 85 on the Stanford-Binet at age 5 years. However, he has a typical Down's syndrome pattern of reasoning in other areas of cognition and by the age of 19 years, the Stanford-Binet decreased to 55 although he still remains in the normal range of vocabulary tests (Penrose, personal communication).

Down's syndrome occurs with a frequency between 1/520 (Hug, 1951) to 1/873 (Parker, 1950) patients per total live births. The most complete survey (Collmann and Stoller, 1962) had an incidence of 1/688 patients per live births. The early mortality formerly seen in Down's syndrome patients is decreasing with the advent of more individualized care and modern cardiac and infectious therapies. Forssman and Akisson (1965) studied 1,263 subjects over 5 years of age with Down's syndrome and found a mortality rate only 6% above normal. They showed that patients who live until 5 years of age are very likely to live past 40 years of age. Almost all older patients live in institutions; these mortality figures are reflected in the cost of 1.5 billion dollars annually for care of Down's syndrome patients (Kramer, 1971).

Down's syndrome fetuses can now be detected by amniocentesis early in pregnancy. If therapeutic abortions of such fetuses become widespread, the incidence of the syndrome will decrease. At this time, the development of a program designed to ameliorate the mental retardation and improve the level of functioning in Down's syndrome patients remains a high priority. Two complementary approaches appear feasible.

The first approach, under extensive investigation by many dedicated professionals throughout the world today, is an attempt to maximize brain function by early environmental stimulation with educational and training techniques. Very early stimulation of the patient and support of the parents to minimize rejection of the infant during the crucial early years of brain development appear to be the key portions of these programs. Individual case reports such as the boy with a 28 I.Q. who could read at a sixth grade level (Butterfield, 1962) suggest these approaches may change the course of future planning for many of these patients.

A second approach involves attempts to compensate for the abnormal biochemical milieu in the central nervous system. There are many objections to this approach, citing the multiple genetic misinformation coming from an entire extra chromosome and the evidence in most of the patients that CNS depression is already present at birth, in spite of maternal placental compensation factors. Yet, even without therapeutic intervention, some Down's syndrome patients do achieve nearly normal intelligence. Also, in contrast to the devastated brains of untreated phenylketonurics (Poser and Van Bogaert, 1959), it is difficult to find agreement among authors that there are consistent pathological abnormalities in the majority of Down's syndrome brains, other than their small size (Crome and Stern, 1967). Solitare and Lamarche (1965) state "there does not appear to be any common denominator which would enable one who has access to the brain of a mongoloid to say indeed, that the particular brain was that of a mongoloid and not of any other individual." Biochemical studies of lipid and other components in the cerebrum also show essentially normal ratios in most of these patients (Stephens and Menkes, 1969). Only recently, using the electron microscope, are consistent differences between normal and Down's syndrome brains finally being reported (Ohara, 1972).

In contrast to pathological studies of the brain, biochemical reports elsewhere in the body have shown many consistent abnormalities in this patient group. Of the many intracellular biochemical abnormalities described, it is reasonable to presume that a certain percentage also apply to the central nervous system. In this monograph, we have explored the effect of changing the level of an amine, serotonin, known to be low in the platelets of these patients. This amine is also present in the human brain. Abnormal (either low or high) blood levels of serotonin (5-hydroxytryptamine, 5-HT) have been reported in many different types of mental retardation (see Chapter 9). It is possible that this finding in the blood in six or more causes of retardation is more than an extraordinary coincidence. Normal functioning of the intellect is interfered with by errors in different metabolic pathways in the brain. Is serotonin part of one such pathway or is it part of a series of complex systems reflecting other metabolic errors? The role of this amine in young patients functioning at sub-optimal brain levels is not understood. Our early studies with administration of the precursor of serotonin suggested that the amine had an effect on muscle tonus in neonatal patients. Studies by other investigators, mostly using animals, suggested a number of other central nervous system functions could be attributed to the amine. An initial effort by my colleagues and myself to explore the role of serotonin in one form of mental retardation, Down's syndrome, is described in this monograph.

MARY COLEMAN, M.D.
(formerly Mary Bazelon)

REFERENCES

BUTTERFIELD, E. C. (1962) A provocative case of over-achievement by a mongoloid. *Am. J. Ment. Def.* **66**, 444.

CARTER, C. H. (1967) Unpredictability of mental development in Down's syndrome. *Southern Med. J.* **60**, 834.

COLLMANN, R. D. and STOLLER, A. (1962) Notes on the epidemiology of mongolism in Victoria, Australia from 1942 to 1957. *Proc. Lond. Conf. Sci. Study Ment. Defic.* **2**, 517.

CROME, L. and STERN, J. (1967) The pathology of mental retardation, L. & A. Churchill Ltd., London.

FINLEY, S. C., FINLEY, W. H., ROSECRANS, C. J. and PHILLIPS, C. (1965) Exceptional intelligence in a mongoloid child of a family with a 13–15/partial 21 (D/partial G) translocation. *New Engl. J. Med.* **272**, 1089.

FORSSMAN, H. and AKISSON, H. O. (1965) Mortality in patients with Down's syndrome. *J. Ment. Def. Res.* **9**, 146.

HUG, E. (1951) Das Geschlechtsverhältnis beim Mongolismus. *Ann. Paediat.* **177**, 31.

HUNT, N. (1967) The world of Nigel Hunt, Garrett Publications, New York.

KRAMER, B. (1971) Diagnoses of fetuses to determine defects spur growing debate, Wall Street Journal, October 14.

LANGDON-DOWN, J. (1866) Observations on an ethnic classification of idiots. *Clin. Lectures and Reports, London Hospital*, **3**, 259.

LEJEUNE, J., GAUTIER, M. and TURPIN, R. (1959) Etudes des chromosomes somatiques de neuf enfants mongoliens. *C.R. Acad. Sci.* **248**, 602.

OHARA, P. T. (1972) Electron-microscopical study of the brain in Down's syndrome. *Brain*, **95**, 681.

PARKER, G. F. (1950) The incidence of mongoloid imbecility in the newborn infant. *J. Pediat.* **36**, 493.

POSER, C. M. and VAN BOGAERT, L. (1959) Neuropathologic observations in phenylketonuria. *Brain*, **82**, 1.

SHAW, M. W. (1962) Familial mongolism. *Cytogenetics*, **1**, 141.

SOLITARE, G. B. and LAMARCHE, J. B. (1965) Alzheimer's disease and senile dementia as seen in mongoloids: neuropathological observations. *Am. J. Ment. Def.* **70**, 840.

STEPHENS, M. C. and MENKES, J. H. (1969) Cerebral lipids in Down's syndrome. *Dev. Med. Child Neurol.* **11**, 346.

STRICKLAND, C. A. (1954) Two mongols of unusually high mental status. *Brit. J. Med. Psychol.* **27**, 80.

ZELLWEGER, H., GROVES, B. M. and ABBO, G. (1968) Trisomy 21 with borderline mental retardation. *Confinia Neurologica*, **30**, 129.

CHAPTER 1

Baseline Serotonin Levels in Down's Syndrome Patients

MARY COLEMAN and DERSH MAHANAND

Introduction

Patient B-15 is a wan little girl with sparse, lemon-colored curls. One notices her gentle smile in spite of an intermittent lateral nystagmus. Because she was a patient in the research study described in this monograph, we have been following her serotonin levels in blood since birth. The child received only placebo during the three-year study. As with most patients with the trisomy 21 form of Down's syndrome, the serotonin levels were significantly lower than those of normal children. During an out-patient visit on December 15, 1970, when the child was 2 years 8 months of age, these levels fell unusually low, even for a Down's syndrome patient. (We measure the levels as total 5-hydroxyindoles—5-HI.) (Fig. 1-1.)

During this particular out-patient visit, we noted that the patient's usually happy mother was very tearful and depressed. We elected to draw a 5-HI level on the mother also, because of the extensive literature summarized by Carpenter (1970), relating low serotonin and clinical depression. The mother's level was about one half the minimum normal adult value that day (Fig. 1-1).

A psychiatric interview disclosed that the patient's mother had become suddenly and severely depressed as the result of an unexpected disclosure by her husband. During the following weeks, the couple resolved their marital crisis and the mother's depression lifted. And so did the mother's 5-HI and the child's 5-HI levels, which rose together. The child's level rose to her former Down's syndrome baseline level; the mother's level rose towards the adult normal (Fig. 1-1).

This monograph is an exploration of serotonin levels in infants and young children with Down's syndrome, who live at home. The story of patient B-15, which suggests an interrelationship between a mother's and child's 5-HI levels, introduces the first of the many complexities of interpretation of these serotonin levels in very young children.

This first chapter surveys current knowledge about serotonin levels in Down's syndrome, including baseline data on 206 patients studied during the last 5 years in our clinic. Daily values on normal children are included for the first time in the literature. The following chapters describe a double blind experiment where the effects of administering the serotonin precursor, 5-hydroxytryptophan, was contrasted with administering placebo in Down's syndrome patients. Specialists independently evaluated serotonin and catecholamine metabolites, muscle tone, temperature, mental functioning, personality patterns and the EEG in these double blind patients. Chapters on other methods of changing serotonin levels and on interpreting serotonin levels in children, including other forms of mental retardation, complete the monograph.

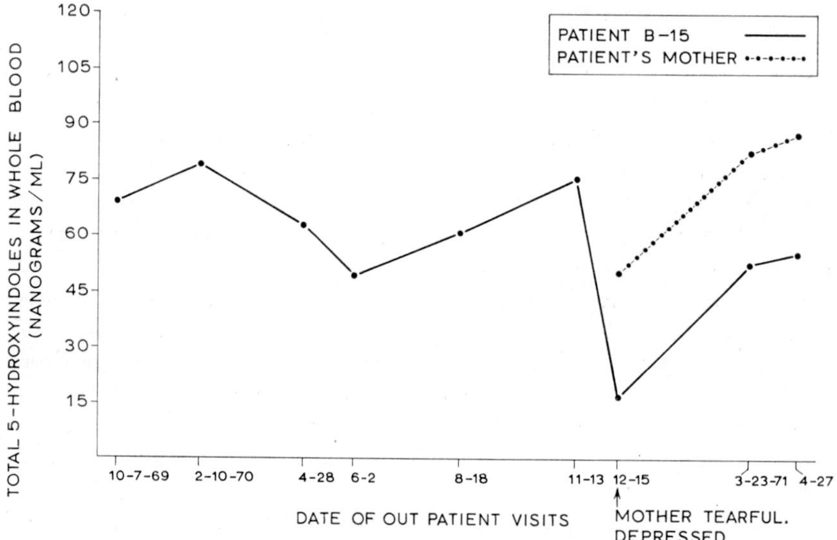

Fig. 1-1. Apparent correlation of a mother's depression with a decrease of 5-HI values in her child. A young patient with trisomy 21 form of Down's syndrome received placebo during the 18 months her blood 5-HI levels are graphed in this figure. The only major decrease in 5-HI values occurred during a period when the patient's mother became clinically depressed. As the mother's mood improved, the 5-HI level of both mother and patient rose.

Serotonin Biochemistry and Methodology

5-HT is present in several different body tissues—the enterochromaffin system of the gastro-intestinal tract, the spleen, the platelets in blood and the central and peripheral nervous systems. Studies by Fuxe and his co-workers (1968) in rats have shown that the 5-HT nerve cell bodies are mainly localized in the raphe nuclei of the lower brainstem; some are also found surrounding the pyramidal tract and in the medioventral part of the caudal tegmentum. No 5-HT cell bodies are found in the spinal cord, the diencephalon or the telencephalon. The axons from the 5-HT nerve cell bodies extend throughout the nervous system with 5-HT terminals concentrated in the visceral nuclei of pons and medulla, reticular forma-

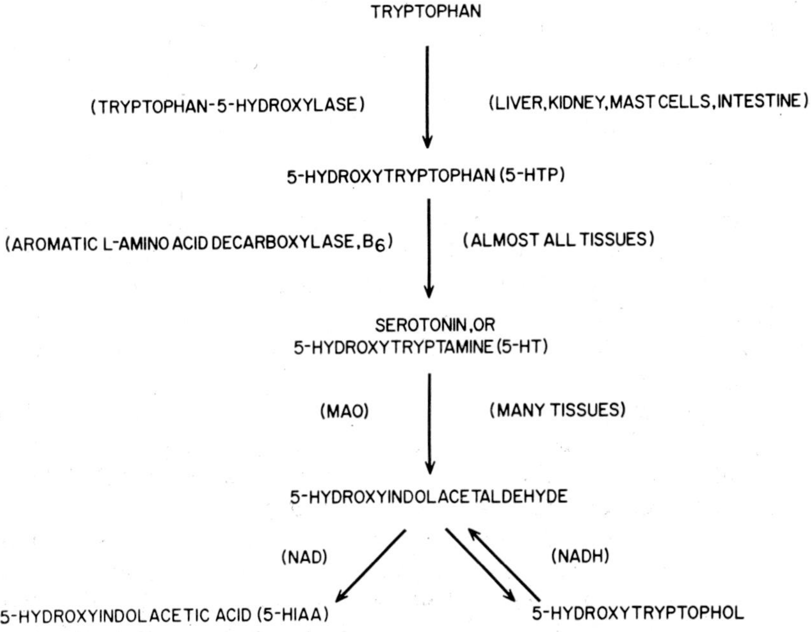

Fig. 1-2. The major metabolic pathway of serotonin in humans.

tion, hypothalamus, thalamus, amygdala and septal area. In the rat, the cerebellum has relatively few 5-HT nerve terminals.

The major metabolic pathway of serotonin (5-hydroxytryptamine [5-HT]) is seen in Fig. 1-2. Only approximately 1% of tryptophan goes to 5-HT. The first step of the pathway, from tryptophan to 5-hydroxytryptophan (5-HTP), is thought to be the rate-limiting one. Under normal conditions, the amount of 5-HT synthesized in brain is controlled by the activity of the first enzyme, tryptophan hydroxylase (Tyce *et al.*, 1967; Jequier *et al.*, 1969), and the tissue levels of tryptophan (Culley *et al.*, 1963) and phenylalanine (Lovenberg *et al.*, 1968). The second step of the pathway uses an ubiquitous enzyme, aromatic L-amino acid decarboxylase, with its co-enzyme, vitamin B6. In newborns and under unusual conditions of co-enzyme loading, this step may have some rate-limiting role.

5-HT is metabolized by monoamine oxidase and aldehyde dehydrogenase to form 5-hydroxyindoleacetic acid (5-HIAA). This is the major end-product of 5-HT in humans. Another important pathway for 5-HT (not shown in the figure) is from N-acetylserotonin to melatonin in the pineal gland.

In this monograph, we measured 5-HT in blood and 5-HIAA in 24-hour urine specimens and most of what we were measuring comes from gastrointestinal production of 5-HT. Haverback and Davidson (1958) studied an adult patient with resection of the large and small intestine and demonstrated that 97% of his whole blood 5-HT and 75% of his urinary 5-HIAA originated from the intestines. The lowest level of 5-HI ever recorded in a patient in our laboratory was approximately 0.5 ng/ml. This newborn died at 7 days of age and autopsy showed agenesis of the gastrointestinal system. Serotonin synthesized outside the central nervous system does not cross into the brain because of selective transport ('blood brain') barriers. CNS 5-HT is made *de novo* from tryptophan. Bertaccini (1960) has shown in the rat the total removal of the gastrointestinal tract has very little effect on brain 5-HT in spite of decreasing serum, spleen and lung 5-HT levels.

There are many methods of measuring serotonin and related 5-hydroxyindoles. These methods are by bioassay (Erspamer, 1966), gas–liquid chromatography (Fales and Pisano, 1962), spectrophotometry (Jepson, 1960) and fluorometric analyses (Bogdanski *et al.*, 1956). For the concentrations of 5-HT present in body fluids, the fluorometric analysis is the most widely accepted method (see review by Lovenberg and Engelman, 1971). The method predominantly used in this study (Appendix I-1) is a fluorometric determination of total 5-hydroxyindoles in whole blood.

When we began this study, we chose a total 5-hydroxyindole (5-HI) rather than the pure serotonin (5-HT) assay for our large number of routine studies, because it was a shorter technique with less chance of error when performed on a mass basis. Total 5-HI is usually 99% 5-HT and 1% 5-HIAA in human whole blood, except for patients with 5-HT, 5-HTP or 5-HIAA producing carcinoid tumors (Sjoerdsma *et al.*, 1957; Oates and Sjoerdsma, 1962) and in patients receiving 5-HTP (Bazelon *et al.*, 1967). In addition to the macro method used throughout this monograph, a micro method suitable for newborns is also included in Appendix I-1. 5-HT, 5-HTP and 5-HIAA assays used the Udenfriend *et al.* (1958) methods.

We also chose a whole blood rather than a platelet assay. The platelets sweep the blood free of 5-HT: virtually all 5-HT is concentrated inside the platelets by an active transport system in the platelet membrane. For this reason, several methods have been devised to estimate the 5-HT in blood by concentrating the platelet fraction and relating the results to platelet count or platelet protein (Lovenberg and Engelman, 1971). These methods give a purer 5-HT assay than the whole blood method which includes any 5-HIAA present in the serum. The platelet methods may be of particular value for assays performed in the first week of life when platelet counts occasionally are quite low in Down's syndrome patients.

However, there are both technical and theoretical problems with these methods. There is some evidence that 5-HT 'leaks out' of human platelets when they are washed during the procedure (Stacey, 1958; Born and Gillson, 1959). Also, surprising as it may seem, 5-HT levels may be one factor influencing the platelet count (Pare *et al.*, 1960; Bazelon *et al.*, 1968). It appears that platelet counts might move in the direction of endogenous levels—high level, high count or low level, low count—decreasing the clinical value of 5-HT methods corrected by platelet counts or protein. Analysis of platelet/5-HT levels in this monograph, although they display a trend in the postulated direction, have failed to substantiate this correlation with statistical significance in our patient sample (see Appendix III-9). Platelet counts were performed on a Model B Coulter Counter manufactured by Coulter Electronics, Inc., Hialeah, Florida.

Statistical significance of results reported in this monograph is determined by standard analytical methods.

5-Hydroxyindoles recorded in this text are based on blood samples taken after a 48-hour, serotonin-low diet, unless noted.

Catecholamine Methodology

The catecholamines (dopamine, norepinephrine and epinephrine) are cousins of serotonin. They share with 5-HT many of their enzymes and co-enzymes, mutually used active transport systems, and partially interchangeable binding sites. Factors affecting the metabolism of 5-HT inevitably partially affect the catecholamine pathways, too; catecholamines and their metabolites in 24-hour urine specimens are determined by the following methods: epinephrine (Kiang *et al.*, 1969), norepinephrine (Kiang *et al.*, 1969), homovanillic acid (HVA) (Goldenberg and Drewes, 1971), vanilmandelic acid (VMA) (Sunderman *et al.*, 1960) and 3-methoxy-4-hydroxyphenylethyleneglycol (MHPG) (Dekirmenjian and Maas, 1970).

All values reported for catecholamines and their metabolites were taken after a 48-hour catecholamine-restricted diet.

First Visit 5-Hydroxyindoles in Patients with Down's Syndrome in our Clinic

Trisomy 21

There are a number of studies of 5-HT metabolism in Down's syndrome patients, but no large series or detailed study of the crucial first year of life. We have studied total 5-hydroxyindoles in 206 children with Down's syndrome.

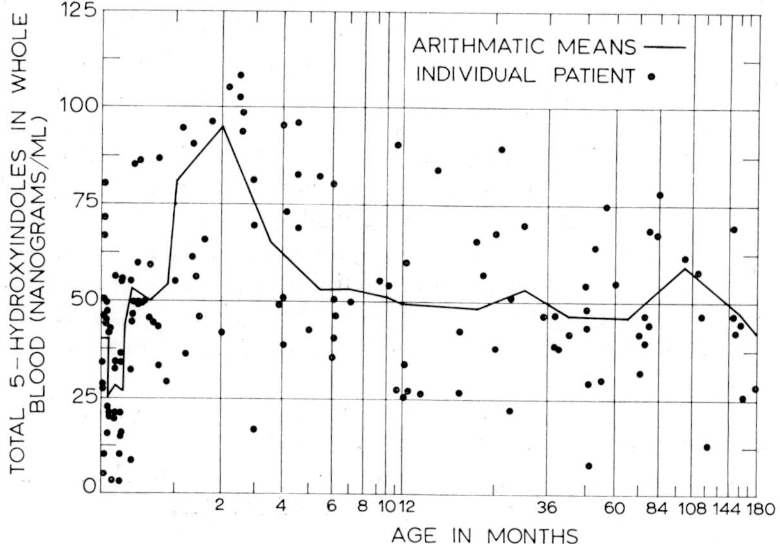

Fig. 1-3. Individual total 5-hydroxyindole (5-HI) values recorded in 174 out-patients with trisomy 21 at the time of their first visit to our research clinic. The arithmetic mean of these values is probably most accurate in the early months of life where larger numbers of patients were tested.

Chromosomal studies disclosed that 174 patients had a trisomy 21 karyotype, 12 had single translocation karyotypes, 12 patients were trisomy 21/normal mosaics, and 8 had other chromosomal variations. In addition, 7 phenotypical mongols with normal chromosomes were studied.

Total 5-HI levels were recorded at the time of the patients' initial visit to the clinic. Later 5-HI values possibly may be affected by many factors, including the supportive atmosphere of the clinic (see Fig. 3-6 in Chapter 3). The first visit values are considered as a baseline sample of children with Down's syndrome who reside at home. All values recorded in this chapter are of patients on neither medication nor placebo of any kind.

Figure 1-3 charts the first visit out-patient 5-hydroxyindole values we have documented on 174 patients with trisomy 21 chromosomal karyotypes (Appendix I-2a). Some patients have normal values for the first 3 months of life but these do not persist. Fibroblast chromosomal culture of the only

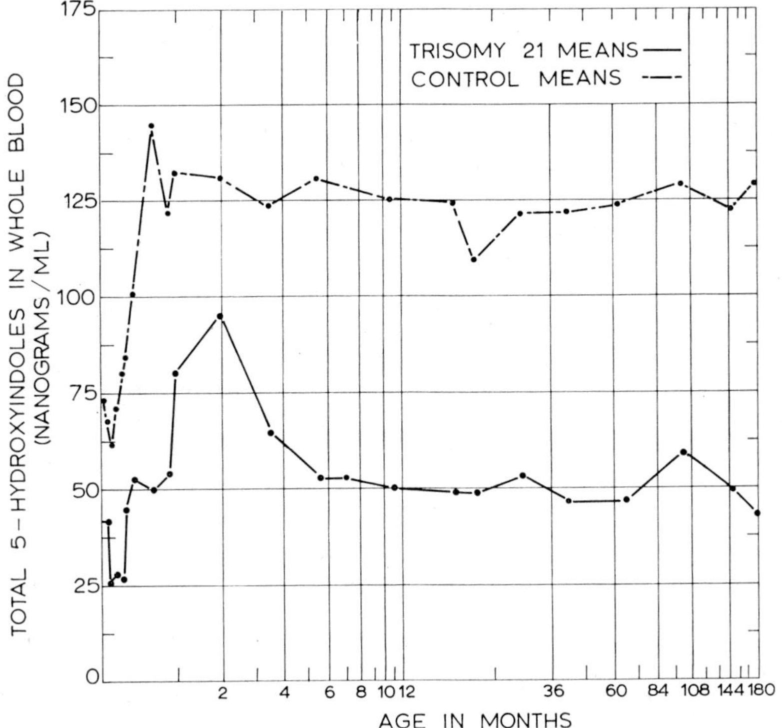

Fig. 1-4. The arithmetic means of total 5-hydroxyindole (5-HI) values in whole blood of 174 patients with trisomy 21 and 174 age-matched controls.

leukocyte trisomy 21 patient with a completely normal 5-HI level after 3 months disclosed a double chromosomal error present only in the fibroblasts (see patient C-33 in Appendix I-5). The comparison of patients and controls in each grouping disclosed that the Down's syndrome patients had 5-HI levels which were statistically significantly lower for every age, except the 2 months category (Appendix I-2a). This is the age group where the mean 5-HI values in the trisomy patients reached their highest level (see Fig. 1-4).

The control patients were infants from normal nurseries and other normal children. However, we excluded some 'normal' children if they had any history or symptoms of behavior patterns that might affect 5-HT levels such as hyperactivity or insomnia (see Chapter 9).

The arithmetic mean of the trisomy cases (Fig. 1-4; Appendix I-2a) shows evidence of an initial decline of 44% in levels from the second to the sixth day of life, possibly due to the loss of maternal compensation factors or recovery from the acute stress of birth (this mimics a similar, less marked, 15% decline in 5-HI values in control patients occurring 48–72 hours after birth). The decline of 5-HI is followed by a gradual increase until a maximum is reached at 2 months of age, again paralleling at a lower and slower level the pattern of control children during this period of infancy.

After 2 months, the values of the trisomy patients rapidly decline to a baseline by 5 months of age of 49 ng/ml which is maintained with minor variations for the next 12 years. The control patients, on the other hand, maintain their 5-HI values at the higher 1-month level of 124 ng/ml from then on. After 5 months of age, this disparity in 5-HI levels between controls and trisomies is greater than in early infancy.

The sample of first visit trisomy 21 patients included all patients referred to our office who had a leukocyte karyotype completed. Most patients were white and 54% were male. The medium income range of the parents of the patients was well above the national average. This series of patients is not, of course, an unbiased sample of Down's syndrome—as in geographical ascertainment studies (Cowie, 1970; Øster, 1953; Hall, 1964). These are patients who survived long enough and had parents and physicians interested enough to refer them to a clinic conducting research on Down's syndrome.

This large clinic study of 5-HI values reported here was the outgrowth of a widespread interest in serotonin in these patients, initially begun in 1960 by Jérôme, who noted low 5-HIAA excretion (Jérôme *et al.*, 1960), and by O'Brien, who found poor response to 5-HTP loading (O'Brien *et al.*, 1960). The first paper to actually examine 5-HT levels in Down's syndrome was by Pare *et al.* (1960). These authors made a major contribution to the literature by pointing out the presence of high 5-HT levels in several forms of retardation (see Chapter 9). However, their studies with the Down's syndrome patients showed normal levels in serum. They studied 10 patients, one of whom had a very high value (?translocation, ?mosaic, ?double chromosomal error). The other 9 patients were somewhat low, but the one extreme case probably brought their mean value up into the low normal range. Another problem was the serum methodology, available at that time, which is a relatively inaccurate method. Two years later Paasonen and Kivalo (1962) published a study comparing the serum and platelet methods of determining 5-HT, using a group of severely retarded patients including six mongols. Their platelet method data show the Down's syndrome patients consistently had lower levels than all other patients studied, so this is the initial paper in the literature to show depression of 5-HT. However, because of the emphasis on the serum method results (no consistent pattern) and lack of normal controls, the authors did not interpret their platelet method results.

In 1965, Rosner and co-authors described low 5-HT levels in Down's syndrome patients using the whole blood method and this finding was quickly confirmed by Tu and Zellweger (1965). Since then, Berman *et al.* (1967), Bazelon *et al.* (1967), McCoy *et al.* (1968), Boullin and O'Brien (1971), Airaksinen (1971) and Lott *et al.* (1971) have also confirmed this result.

In our clinic, we had our most complete data for the neonatal period. Animal studies have established that 5-HT is low in the central nervous system of many newborn animals (Tissari and Pekkarinen, 1966) probably due to low level of the first enzyme in the 5-HT pathway—tryptophan hydroxylase—(Bennett and Giarman, 1964) although immature binding of the amine, co-factor availability and presence of isoenzymes may be other relevant factors (Eiduson, 1971). There are several studies showing the same pattern of relatively low levels in the blood of normal human newborns (Mitchell and Cass, 1959; Hazra *et al.*, 1965; Berman *et al.*, 1965). The Berman *et al.* (1965) study suggests that immaturity of the second decarboxylase step of the 5-HT pathway plus relative deficiency of its co-enzyme vitamin B6, may be a factor in low 5-HT levels in human newborns. Berman *et al.* (1965) also has the only serial study in the first two weeks of life in normal neonates and he showed a continuous rise every 3 day period. By examining levels in 24-hour segments in our clinic, we saw a post-birth dip lowest at 48–72 hours in normal children. Hazra studied 5 selected periods up to 24 weeks in normal children and found the highest 5-HT level at 4–6 weeks. In our normal patients, our peak 5-HI value was at 15–21 days. The trisomies lagged behind but moved toward the normal patterns, behaving more like low birth weight infants (Berman *et al.*, 1965). We saw no statistically significant sexual difference in 5-HI values among our young controls and patients, perhaps because of limited numbers, as have been described in rats (Ladosky and Gaziri, 1970).

We had relatively few patients and controls in the later childhood ages, and did not see the gradual decline in 5-HI values up until adolescent years reported by Ritvo *et al.* (1971) in normals. In this fine study, all 5-HI assays were begun within an hour of the time of blood collection, avoiding the loss of 5-HT in blood samples with time.

Mosaics

First visit 5-HI values were obtained in 12 Down's syndrome patients with trisomy 21/normal leukocyte karyotypes (Appendix I-3). One of these children.

aged 10 years, with a 50% mosaicism (46,XY/47,XY, G+), had a 5-HI value of 150 ng/ml (see Fig. 1-5). A repeat value to check this unusual value was 169 ng/ml, corroborating the initial results. On the Stanford–Binet Intelligence Scale (L-M), this patient's I.Q. was 35. It was difficult to test him due to his severe hyperactivity, a characteristic that distinguished him from the majority of Down's syndrome patients.

If a calculation of the arithmetic mean includes the patient with the very high level, this brings the mean of cases over 6 months of age to 74 ng/ml, while excluding this one case yields a value of 55 ng/ml (see Appendix I-4 for exact figures). Although these means are slightly higher than those determined for the trisomy patients of comparable age, a statistically significant difference of the arithmetic mean (excluding the high case) is not present.

Also, with the small number of cases we had, the percentage of abnormal cells in the leukocytic karyotype at the time of the 5-HI determination does not appear to be related to this level, *i.e.*, a higher percentage of abnormal cells does not necessarily correlate with a more abnormal 5-HI level. All but two of our patients were under 3 years of age, a period where rapid cell selection is changing the percentage of mosaicism in the lymphocytes (Taylor, 1970). No study of levels of 5-HI in children with the mosaic form of Down's syndrome has been previously reported. In our 1967 paper (Bazelon *et al.*, 1967) we reported the 5-HI level of one child only. Although in the course of this present study we found one mosaic patient with a high normal level of 5-HI (just as Berman *et al.* (1967) have reported one translocation with a high level), there is no statistical evidence that mosaics have better 5-HI levels based on our limited series of 12 patients.

Translocations

First visit 5-HI levels were obtained on 12 patients with the translocation karyotypes and clinical Down's syndrome (see Fig. 1-5). A comparison with trisomy means shows no significant difference (Appendix I-4). The arithmetic mean for the three patients over 6 months of age was 33 ng/ml. Because of the paucity of these older cases, it is impossible to say whether this is significantly below the mean of the trisomies for this age group (49 ng/ml).

In the case of translocations, there have been two previous 5-HT studies reported. The Rosner *et al.* paper (1965), an early report from our laboratory of 7 translocations, found these patients had 5-HI levels which were 81% of normal, compared to the trisomies who were 49% of normal. Berman *et al.* (1967) studied 9 translocations and found the levels about 66% of normal. However, he had one patient (a D/G translocation) with a very high (415 ng/ml) value. Omitting his one high value, his mean percentage drops to 45%, below his trisomy mean.

Down's Syndrome Patients with Two Chromosomal Errors

One of the interesting findings of this study was the relatively high 5-HI levels seen in 2 patients with clinical Down's syndrome and a standard chromosomal finding combined with a second chromosomal error (Table 1-1). Many combinations of chromosomal errors have been reported in Down's syndrome patients (Tsuboi *et al.*, 1968). The highest first visit value recorded in a patient with two chromosomal errors (148 ng/ml) was clearly significant since it was

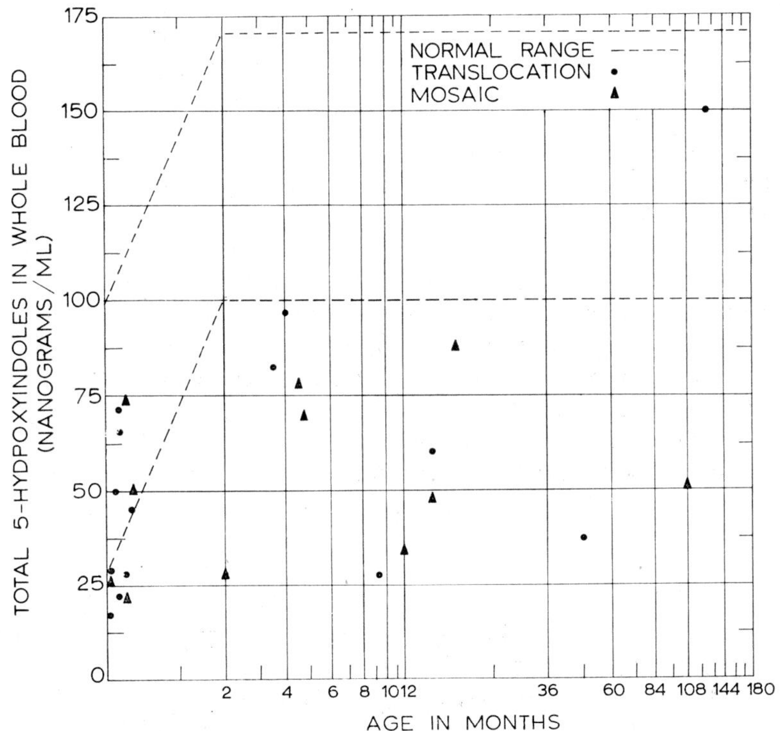

Fig. 1-5. Individual first visit 5-HI values in patients with the mosaic (46,XY/47,XY,G+) and translocation forms of Down's syndrome.

well above any trisomy values recorded for those ages. This patient (C-33) had a double error not present in the leukocytes. His fibroblasts were cultured because of his high 5-HI level and the second error was then discovered. The other double chromosomal error patient (C-30), also had a quite high 5-HI value, well above the trisomy mean and slightly above the highest pure trisomy value for age.

The neurological examination of these patients, particularly the onus

TABLE 1-1

First visit total 5-hydroxyindole values in two patients with double chromosomal errors

Patient	Leukocytic karyotype		5-Hydroxy-indoles		Comparison with trisomy mean for age	
	One error	Second error	Age	Level (ng/ml)	Trisomy mean (ng/ml)	Nano-gram increase
C-30	47,XY,21+	46,XY,t(Dp^-,Dq^+)	4 yrs	94	53	+41
C-33	47,XY,21+	48,XXY,21+*	10 mths	148	58	+90

* Leukocytic karyotype showed only 47,XY,21+ ; the fibroblast karyotype revealed the double chromosomal error.

scores, was relatively good compared to the presenting examination in most Down's syndrome patients. This phenomenon of relatively normal neurological findings and double chromosomal errors has already been observed in the literature. Cowie studied 79 Down's syndrome children neurologically and reported the least hypotonic and most alert in her entire series was the only patient with a double chromosomal error (trisomy 21 and balanced B/C translocation). She states "it is open to question whether the balanced translocation in this case has conferred a biological advance" (Cowie, 1970).

Phenotypical 'Down's Syndrome' Patients—Pseudomongols and Paramongols

Seven patients were referred to our clinic who had stigmata of Down's syndrome but normal leukocytic and fibroblastic chromosomes. It is, of course, possible that these children may have had abnormal chromosomes in other tissues not easily accessible to study; terms such as hidden mosaics or *forme fruste* have been used to describe them. Hall (1962) has chosen the name 'pseudomongoloids' for these patients if they are mentally retarded. The term 'paramongoloid' applies to those with normal intelligence (Appendix I-5). Characteristics which help distinguish these children from patients with chromosomal errors are (1) less than six clinical stigmata designated as cardinal signs (Hall, 1964), (2) an occipito-frontal head circumference in excess of the 45th percentile, and (3) 5-HI levels borderline or normal. When initially seen in the newborn period, these patients present a serious clinical problem and chromosomal analysis is often indicated. The term 'Down's syndrome' in this monograph is limited to patients with demonstrable chromosomal errors.

Discussion

Why is 5-HT low in the platelets in patients with Down's syndrome? There are a number of possibilities and studies to support many of these. In a disease entity with so many genetic errors as trisomy 21, more than one factor may be involved. This is especially likely regarding this particular amine—5-HT—which is affected by such a variety of factors in other diseases (see Chapter 9). In chromosomal disease there is excessive, *but originally normal*, genetic information available in each cell. Interpretations of data which lead back to overproduction by genes, including those regulatory or inhibitory, seem most likely to be productive.

The first step of the 5-HT pathway (Fig. 1-2) is the absorption of tryptophan consumed in food from the gastrointestinal tract, which contains fewer enterochromaffin cells (serotonin-making cells) in Down's patients (Couturier-Turpin *et al.*, 1971). O'Brien and Groshek (1962), have shown that blood tryptophan levels are lower in Down's syndrome and the studies of Jérôme and his colleagues (1960, 1962) suggest that intestinal absorption of administered L-tryptophan is less efficient. They also postulated an acceleration in the kynurenine pathway (another tryptophan pathway) leading to a secondary effect on 5-HT levels (Jérôme *et al.*, 1960). Tissot *et al.* (1966), also showed a significant increase in 5-HIAA with tryptophan loading in mongols compared to other retardates. Because of the lower concentration of albumin, Airaksinen and Airaksinen (1972) report that the binding of tryptophan is lowered in plasma. Since availability of tryptophan is one rate-limiting factor determining the endogenous level of 5-HT (Culley *et al.*, 1963; Fernstrom and Wurtman, 1971), low levels of tryptophan may be a factor in the low 5-HT seen in Down's syndrome.

Another theoretical possibility (not yet explored in quantitative detail in the literature) is the possibility of inhibition of the active transport systems moving tryptophan from the gastrointestinal system and into the central nervous system by excessive amounts of other large neutral amino acids who share the same transport mechanism.

Tryptophan is the natural precursor of 5-HT in the brain. It is converted to 5-HTP by tryptophan hydroxylase, which in most instances is the rate-limiting enzyme of the pathway. Inhibition of this enzyme for any reason, such as overstimulation of controlling feedback pathways, or overuse of its co-enzyme by other metabolic pathways, would definitely limit the amount of 5-HT produced. These studies, difficult to perform on patients, have not been done.

5-HTP is then converted into the active amine—5-HT—by the enzyme, L-aromatic amino acid decarboxylase, using vitamin B6 as a co-factor. Although this enzyme is used by many pathways, it has a high affinity for 5-HTP, which partially protects the 5-HT pathway in the presence of competing substrates. In any event, this step of the pathway is not usually rate-limiting, although large dosages raise platelet 5-HI levels in patients with Down's syndrome and other disease entities (see Chapter 8). Some studies on vitamin B6 levels in Down's syndrome, reviewed in detail in Chapter 8, suggest a functional or relative deficiency of this co-enzyme can be demonstrated in these patients. We have not been able to confirm this (Bhagavan *et al.*, 1973).

Another enzyme that might change the endogenous level of 5-HT is monoamine oxidase (MAO) which breaks down 5-HT by oxidative deamination (Paasonen *et al.*, 1964). In rat brains, under normal conditions, Grahame-Smith (1971) has suggested that the rate of brain 5-HT synthesis is normally in excess of that required to fulfil the functional needs of the brain and that one way the excess is controlled is by MAO. It can be postulated that very increased amounts of this enzyme might accelerate breakdown and lower the endogenous level of the amine below normal. However, apparently this is not a factor in the platelets of Down's syndrome patients since careful studies by Benson and Southgate (1971) have shown the levels in the platelets to be significantly lower, not higher, than controls. Also, Boullin and O'Brien (1971) have reported a decrease in the formation of 5-HT metabolites by platelet monoamine oxidase in this patient group.

However, in spite of these many studies showing deficiencies at each step of the metabolic pathway, there is evidence that this pathway may be essentially intact in these patients. A comprehensive study of the functioning of the steps of the 5-HT pathway all at the same time was performed by Engelman. He injected ^{14}C-labeled tryptophan into four children with trisomy 21 and measured ^{14}C-5-HIAA in the urine. His results did not differ from studies on normal patients (Engelman *et al.*, 1970). The only factor not studied by such a procedure is intestinal absorption of tryptophan, already reported to be diminished in this patient group. The Engelman study suggests that these patients, a group known to have greatly diminished 5-HT levels in platelets, may be making enough 5-HT to keep those platelets normally full. This suggests that the platelets must be failing to take it up or keep it properly. Other evidence of normal 5-HT metabolism in Down's patients was shown by Perry (1962) who found no difference in urinary secretion of 5-HT in patients and controls. Also, we have demonstrated that five times the normal amounts of 5-HT can be produced in the blood by a trisomy 21 patient with high loading dosages of 5-HTP (Bazelon *et al.*, 1968). There may be no major impairment in ability of these patients to synthesize 5-HT in the body generally.

Regarding the central nervous system, several studies have suggested normal synthesis of 5-HT is present there also. Normal levels of 5-HIAA have been measured in the cerebral spinal fluid of these patients (Dubowitz and Rogers, 1969); however, Fallstrom *et al.* (1973) reported lower levels. Tu and Partington (1972) and Lott *et al.* (1971) estimated central turnover of 5-HT by probenicid studies of 5-HIAA in cerebral spinal fluid and were satisfied that they had demonstrated normal rates of turnover in Down's syndrome patients. Thus we have some evidence of normal 5-HT turnover in the body in the presence of greatly diminished platelet levels. We also have evidence of normal 5-HT turnover in the brain; does it also exist in the presence of greatly diminished serotonergic neurone levels? (See Table 1-2.)

TABLE 1-2

Comparison of 5-HT turnover and endogenous levels in Down's syndrome patients

	Systemic	CNS
5-HT turnover	normal	normal
Endogenous 5-HT level	Markedly decreased in platelets	? In serotonergic neuron

In Down's syndrome, binding may be more crucial than synthesis in regulating the amount of 5-HT available at functional sites.

If there is a normal amount of 5-HT made, why do the platelets have such diminished amounts? It is likely that uptake and binding factors may provide a major part of the answer regarding low 5-HT in platelets in Down's syndrome patients. The binding sites for 5-HT are similar in both the platelet and the neurone, where binding assures adequate delivery of amines to their transmitter sites. This concept that binding may be as crucial as synthesis in regulating the amount of 5-HT available for functional activity is not a new one (reviewed by Grahame-Smith, 1971).

Normally, 5-HT (mostly from the enterochromaffin system of the gastrointestinal tract) is swept out of the blood by the platelet, entering through an energy-dependent transport mechanism in the platelet membrane. 5-HT is then stored as a high molecular weight aggregate with adenosine triphosphate (ATP) (Berneis *et al.*, 1969) in electron-dense subcellular organelles of platelets (Davis and White, 1968).

There have been four studies of 5-HT uptake into the platelets of Down's syndrome patients. Jérôme and Kamoun (1967), Boullin and O'Brien (1971), McCoy and Bayer (1973), and Lott *et al.* (1972) found the uptake or transport into the platelets to be depressed. Other transport systems, such as the rates of flux of sodium and potassium across the red cell membrane, have been reported as abnormal (Baar and Gordon, 1963).

After entering the platelet, a micelle of 5-HT and ATP is formed in the storage organelles. Is there enough ATP in mongol platelets to bind 5-HT?

Studies have shown opposite results. Jérôme and Kamoun (1967) reported normal concentrations of ATP in the platelets of these patients; Boullin and O'Brien (1971) reported a reduction of ATP to only 26% of normal levels, and McCoy and Bayer (1973) confirmed it. Further studies are needed to clarify this point. If ATP is, in fact, depressed, it might be one important factor in explaining the increased efflux of 5-HT from Down's syndrome platelets reported from our laboratories (Boullin *et al.*, 1969). Failure of adequate drug binding has also been reported in Down's syndrome (Ebadi, 1970). Lott *et al.* (1972) have not confirmed increased 5-HT efflux studies. In these patients, ions and other factors regulating the binding need extensive intracellular studies.

Finally, a careful look at platelet kinetics in Down's syndrome patients is needed. A number of enzyme abnormalities reported in these patients' leukocytes appear to be related, in part, to Down's syndromes' unusual leukocyte turnover rather than to the enzyme level itself. Three enzymes, other than monoamine oxidase, have been investigated in the platelets in Down's syndrome patients—alkaline phosphatase, glucose-6-phosphate dehydrogenase (G-6-PD) (Shih and Hsia, 1966) and phosphofructokinase (Doery *et al.*, 1968)—and all have been found to be normal. Two of these enzymes, alkaline phosphatase and G-6-PD, are elevated in Down's syndrome leukocytes, a blood cell with abnormal kinetics. The same enzymes are normal in platelets. Thus, evidence to date suggests that 5-HT and one of its enzymes (MAO) appear to be abnormal in the platelet, while other tested enzyme systems appear to be intact in this cell.

In conclusion, it is impossible at this time to state definitely which errors result in low 5-HT in whole blood in Down's syndrome. Failure of adequate tryptophan absorption may be a factor, although it is probably not the major one. In our opinion, the most likely explanation of the profoundly low levels seen in this patient group is a failure of platelet uptake and binding to adequately obtain and conserve the possibly normal or close to normal amount of 5-HT available. Whether the disabilities of the platelets in Down's syndrome children are shared by their serotonergic neurons remains a fascinating but unanswerable speculation.

Summary

A compilation of available data in the literature from a few small studies on serotonin levels in Down's syndrome is presented and compared to our series —a larger study where daily 5-HI levels are available in the neonatal period in both patients or controls for the first time. First visit total 5-hydroxyindoles in whole blood were recorded in 174 out-patients with trisomy 21 and 174 age-matched controls between 1 day and 26 years of age. The control patients were selected with an effort to omit those with a history of factors affecting 5-HT levels in apparently normal children; one such factor (a mother's depression lowering her child's 5-HI level) is introduced in this chapter.

Our study shows that 5-HI levels are approximately half of childhood values at birth in normal newborns and decrease by 15% between 36 and 48 hours of age before starting a rise to baseline levels at 1 month of age. Patients with trisomy 21 mimic the early portion of the normal curve at a lower and later level. They start at a much lower 5-HI level at birth and decrease by as much as 44% for a period up to 144 hours before starting the neonatal rise. It takes the patients twice as long (2 months) to reach their maximum level but this level does not hold throughout childhood as in normal controls. Instead, the 5-HI level begins to fall at between 2 and 5 months of age to a new low baseline which persists throughout childhood. The 5-HI baseline past 5 months of age in trisomy 21 patients (49 ng/ml) is 40% of the control children's baseline (124 ng/ml) in this series.

Total 5-hydroxyindoles levels in 12 mosaics and 12 translocation patients were not significantly different from the trisomy 21 levels. 5-HI levels superior to trisomy values were seen in 2 patients with chromosomal variations and 7 phenotypical mongols.

The possible causes of low serotonin levels in platelets in Down's syndrome patients are reviewed. Although failure of adequate tryptophan absorption may be one factor, failure of adequate platelet uptake and binding of 5-HT is more likely responsible for the profoundly low levels consistently recorded in these patients.

REFERENCES

AIRAKSINEN, E. M. (1971) Platelet-rich plasma 5-hydroxytryptamine, urinary 5-hydroxyindoleacetic acid and tryptophan ingestion in mongols. *J. ment. Defic. Res.* **15**, 244.

AIRAKSINEN, E. M. and AIRAKSINEN, M. M. (1972) The binding of tryptophan to plasma proteins and the rate of the inactivation of 5-HT released from platelets in Down's syndrome. *Ann. Clin. Res.* **4**, 361.

BAAR, H. S. and GORDON, M. (1963) Kation fluxes in erythrocytes of mongoloids. In: *Proceedings, II International Congress on Mental Retardation, Vienna*, p. 373.

BAZELON, M., BARNET, A., LODGE, A. and SHELBURNE, S. A. (1968) The effect of high doses of 5-hydroxytryptophan on a patient with trisomy 21. *Brain Res.* **11**, 397.

BAZELON, M., PAINE, R. S., COWIE, V. A., HUNT, P., HOUCK, J. C. and MAHANAND, D. (1967) Reversal of hypotonia in infants with Down's syndrome by administration of 5-hydroxytryptophan. *Lancet*, **1**, 1130.

BENNETT, D. S. and GIARMAN, N. J. (1965) Schedule of appearance of 5-hydroxytryptamine (serotonin) and associated enzymes in the developing rat brain. *J. Neurochem.* **12**, 911.

BENSON, P. F. and SOUTHGATE, J. (1971) Diminished activity of platelet monoamine oxidase in Down's syndrome. *Am. J. Human Genetics*, **23**, 211.

BERMAN, J. L., HULTÉN, M. and LINDSTEN, J. (1967) Blood-serotonin in Down's syndrome. *Lancet*, **1**, 730.

BERMAN, J. L., JUSTICE, P. and HSIA, D. Y.-Y. (1965) The metabolism of 5-hydroxytryptamine (serotonin) in the newborn. *J. Pediat.* **67**, 603.

BERNEIS, K. H., DAPRADA, M. and PLETSCHER, A. (1969) Micelle formation between 5-hydroxytryptamine and adenosine triphosphate in platelet storage organelles. *Science*, **165**, 913.

BERTACCINI, G. (1960) Tissue serotonin and urinary 5-HIAA after partial or total removal of gastrointestinal tract in the rat. *J. Physiol.* (*London*), **153**, 239.

BHAGAVAN, H. N., COLEMAN, M., COURSIN, D. B. and ROSENFELD, P. (1973) Pyridoxical 5-phosphate levels in whole blood in home-reared patients with trisomy 21. *Lancet*, **1**, 889.

BOGDANSKI, D. F., PLETSCHER, A., BRODIE, B. B. and UDENFRIEND, S. (1956) Identification and assay of serotonin in brain. *J. Pharmac. Exp. Ther.* **117**, 82.

BORN, G. V. R. and GILLSON, R. E. (1959) Studies on the uptake of 5-hydroxytryptamine by blood platelets. *J. Physiol.* (*London*), **146**, 472.

BOULLIN, D. J., COLEMAN, M. and O'BRIEN, R. A. (1969) Defective binding of 5-HT by blood platelets from children with the trisomy 21 form of Down's syndrome. *J. Physiol.* (*London*), **204**, 128p.

BOULLIN, D. J. and O'BRIEN, R. A. (1971) Abnormalities of 5-hydroxytryptamine uptake and binding by blood platelets from children with Down's syndrome. *J. Physiol.* **212**, 287.

CARPENTER, W. T. (1970) in: Serotonin now: clinical implications of inhibiting its synthesis with parachlorophenylalanine. *Ann. Int. Med.* **73**, 607.

COUTURIER-TURPIN, M., TAMBOISE, E., COUTURIER, D. and ROZÉ, C. (1971) Cellules entérochromaffines à sérotonine et mongolisme. *C. R. Acad. Sci.* (*Paris*) 540.

COWIE, V. A. (1970) A study of the early development of mongols, Pergamon Press, London, p. 12.

CULLEY, W. J., SAUNDERS, R. N., MERTZ, E. T., JOLLY, D. H. and CORLEY, R. C. (1963) Effect of a tryptophan deficient diet on brain serotonin and plasma tryptophan level. *Proc. Soc. Exp. Biol. Med.* **113**, 645.

DAVIS, R. B. and WHITE, J. G. (1968) Localization of 5-hydroxytryptamine in blood platelets: autoradiographic and ultrastructural study. *Brit. J. Haemat.* **15**, 93.

DEKIRMENJIAN, H. and MAAS, J. W. (1970) An improved procedure of 3-methoxy-4-hydroxyphenylethylene glycol determination by gas–liquid chromatography. *Anal. Biochem.* **35**, 113.

DOERY, J. C. G., HIRSH, J., GARSON, O. M. and GRUCHY, G. C. (1968) Platelet-phosphohexokinase levels in Down's syndrome. *Lancet*, **2**, 894.

DUBOWITZ, V. and ROGERS, K. J. (1969) 5-Hydroxyindoles in the cerebrospinal fluid of infants with Down's syndrome and muscle hypotonia. *Develop. Med. Child Neurol.* **11**, 730.

EBADI, M. S. (1970) Increase in brain pyridoxal phosphate by chlorpromazine. *Pharmacology* (*Basel*) **3**, 97.

EIDUSON, S. (1971) Biogenic amines in the developing brain. *U.C.L.A. Forum in Medical Sciences*, **14**, 391.

ENGELMAN, K. In: SJOERDSMA, A., LOVENBERG, W., ENGELMAN, K., CARPENTER, W. T., WYATT, R. J. and GESSA, G. L. (1970) Serotonin now: clinical implications of inhibiting its synthesis with para-chlorophenylalanine. *Ann. Int. Med.* **73**, 607.

ERSPAMER, V. (1966) 5-Hydroxytryptamine and related indolealkylamines, Springer-Verlag, New York.

ESSMAN, W. B. (1970) Some neurochemical correlates of altered memory consolidation. *Trans. N.Y. Acad. Sci.* **32**, 948.

FALES, H. M. and PISANO, J. J. (1962) Gas chromatography of biochemically important amines. *Anal. Biochem.* **3**, 337.

FALLSTROM, P., LIEDHOLM, M. and LUNDBORG, P. (1973) Evidence of altered cerebral serotonin metabolism in Down's syndrome from measurements of cerebrospinal fluid acids. *Pediat. Res.* 7, 53.

FERNSTROM, J. D. and WURTMAN, R. J. (1971) Brain serotonin content: physiological dependence on plasma tryptophan levels. *Science*, **173**, 149.

FUXE, K., HÖKFELT, T. and UNGERSTEDT, U. (1968) Localization of indolealkylamines in CNS. *Advances Pharmacol.* **6A**, 235.

GIARMAN, J. H. and FREEDMAN, D. X. (1965) Biochemical aspects of the actions of psychotomimetic drugs. *Pharmac. Rev.* **17**, 1.

GOLDENBERG, H. and DREWES, P. A. (1971) Direct photometric determination of globulin in serum. *Clin. Chem.* **17**, 358.

GRAHAME-SMITH, D. G. (1971) Studies *in vivo* on the relationship between brain tryptophan, brain 5-HT synthesis and hyperactivity in rats treated with a monoamine oxidase inhibitor and L-tryptophan. *J. Neurochem.* **18**, 1053.

HALL, B. (1962) Down's syndrome (mongolism) with normal chromosomes. *Lancet*, **2**, 1026.

HALL, B. (1964) Mongolism in newborns: a clinical and cytogenetic study. *Acta Paediat.* (*Uppsala*), Suppl. **154**.

HAVERBACK, B. J. and DAVIDSON, J. D. (1958) Serotonin and the gastrointestinal tract. *Gastroenterology*, **35**, 570.

HAZRA, M., BENSON, S. and SANDLER, M. (1965) Blood 5-hydroxytryptamine levels in the newborn. *Arch. Dis. Child.* **40**, 513.

JEPSON, J. B. (1960) Chromatographic and electrophoretic techniques, Vol. 1, (ed.) I. Smith, Wiley-Interscience, New York, p. 183.

JEQUIER, E., ROBINSON, D. S., LOVENBERG, W. (1969) Further studies on tryptophan hydroxylase in rat brainstem and beef pineal. *Biochem. Pharmacol.* **18**, 1071.

JÉRÔME, H. (1962) Anomalies du métabolisme du tryptophane dans la maladie mongolienne. *Bull. Soc. Med. Hop. Paris*, **113**, 168.

JÉRÔME, H. and KAMOUN, P. (1967) Diminution du taux de la sérotonine associée à un défaut de captation dans les plaquettes sanguines des sujets trisomiques 21. *C. R. Acad. Sci.* (*Paris*) **264**, 2072.

JÉRÔME, H., LEJEUNE, J. and TURPIN, R. (1960) Study of the urinary excretion of some tryptophan metabolites in mongoloid children. *C. R. Acad. Sci.* (*Paris*), **251**, 474.

KIANG, J. J. (1969) Differential analysis of adrenaline and noradrenaline, Biorad Laboratories, Richmond, California.

LADOSKY, W. and GAZIRI, L. C. J. (1970) Brain serotonin and sexual differentiation of the nervous system. *Neuroendocrinology*, **6**, 168.

LOTT, I. T., CHASE, T. N. and MURPHY, D. L. (1972) Down's syndrome: transport, storage, and metabolism of serotonin in blood platelets. *Pediat. Res.* **6**, 730.

LOTT, I. T., MURPHY, D. L. and CHASE, T. N. (1971) Down's syndrome: cerebral monoamine turnover in patients with diminished platelet serotonin. *Neurology*, **21**, 441.

LOVENBERG, W. and ENGELMAN, K. (1971) Assay of serotonin, related metabolites, and enzymes. In: Methods of biochemical analysis, Wiley-Interscience, New York, p. 1.

LOVENBERG, W., JEQUIER, E. and SJOERDSMA, A. (1968) Tryptophan hydroxylation in mammalian systems. *Advances Pharmacol.* **6**, 21.

MCCOY, E. E. and BAYER, S. M. (1973) Decreased serotonin uptake and ATPase activity in platelets from Down's syndrome patients. *Clin. Res.* **21**, 304.

MCCOY, E. E., ROSTAFINSKY, M. J. and FISHBURN, C. (1968) The concentration of serotonin by platelets in Down's syndrome. *J. Ment. Defic. Res.* **12**, 18.

MITCHELL, R. G. and CASS, R. (1959) Histamine and 5-hydroxytryptamine in the blood of infants and children. *J. Clin. Invest.* **38**, 595.

OATES, J. A. and SJOERDSMA, A. (1962) Unique syndrome associated with secretion of 5-hydroxytryptophan by metastatic gastric carcinoids. *Amer. J. Med.* **32**, 333.

O'BRIEN, D. and GROSHEK, A. (1962) The abnormality of tryptophane metabolism in children with mongolism. *Arch. Dis. Child.* **37**, 17.

O'BRIEN, D., HAAKE, M. W. and BRAID, B. (1960) Atropine sensitivity and serotonin in mongolism. *Amer. J. Dis. Child.* **100**, 873.

ØSTER, J. (1953) Mongolism. A clinicogenealogical investigation comprising 526 mongols living on Seeland and neighbouring islands in Denmark, Danish Science Press, Ltd., Copenhagen.

PAASONEN, M. K. and KIVALO, E. (1962) The inactivation of 5-hydroxytryptamine by blood platelets in mental deficiency with elevated serum 5-hydroxytryptamine. *Psychopharmacologia* (*Basel*) **3**, 188.

PAASONEN, M. K., SOLATUNTURI, E. and KIVALO, E. (1964) Monoamine oxidase activity of blood platelets and their ability to store 5-hydroxytryptamine in some mental deficiencies. *Psychopharmacologia* (*Basel*) **6**, 120.

PARE, C. M. B., SANDLER, M. and STACEY, R. S. (1960) 5-Hydroxyindoles in mental deficiency. *J. Neurol. Neurosurg. Psychiat.* **23**, 341.

PERRY, T. L. (1962) Urinary excretion of amines in phenylketonuria and mongolism. *Science*, **136**, 879.

RITVO, E., YUWILER, A., GELLER, E., PLOTKIN, S., MASON, A. and SAEGER, K. (1971) Maturational changes in blood serotonin levels and platelet counts. *Biochemical Medicine*, **5**, 90.

ROSNER, F., ONG, B. H., PAINE, R. S. and MAHANAND, D. (1965) Blood-serotonin activity in trisomic and translocation Down's syndrome. *Lancet*, **1**, 1191.

SHIH, L. Y. and HSIA, D. Y. Y. (1966) Enzymes in Down's syndrome. *Lancet*, **1**, 155.

SJOERDSMA, A., WEISSBACH, H., TERRY, L. L. and UDENFRIEND, S. (1957) Further observations on patients with malignant carcinoid. *Amer. J. Med.* **23**, 5.

STACEY, R. S. (1958) Symposium on 5-hydroxytryptamine, (ed) G. P. Lewis, Pergamon Press, London, p. 125.

SUNDERMAN JR, F. W., CLEAVELAND, P. D., LAW, N. C. and SUNDERMAN, F. W. (1960) A method for the determination of 3-methoxy-4-hydroxymandelic acid ("vanilmandelic acid") for the diagnosis of pheochromocytoma. *Amer. J. Clin. Path.* **34**, 293.

TAYLOR, A. I. (1970) Further observations of cell selection *in vivo* in normal/G trisomic mosaics. *Nature* (*London*), **227**, 163.

TISSARI, A. and PEKKARINEN, E. M. (1966) 5-Hydroxyindoleacetic acid in the developing brain. *Acta Physiol. Scand.* Suppl. **277**.

TISSOT, R., GUGGISBERG, M., CONSTANTINIDIS, J. and BETTSCHART, W. (1966) Excrétion urinaire de l'acide 5-hydroxyindolacétique, de l'acide vanillomandélique et de l'acide xanthurénique chez 18 mongoliens. *Path. Biol.* (*Paris*), **14**, 312.

TSUBOI, T., INOUYE, E. and KAMIDE, H. (1968) Chromosomal mosaicism in two Japanese children with Down's syndrome. *J. Ment. Defic. Res.* **12**, 162.

TU, J. and PARTINGTON, M. W. (1972) 5-Hydroxyindole levels in the blood and CSF in Down's syndrome, phenylketonuria and severe mental retardation. *Develop. Med. Child Neurol.* **14**, 457.

TU, J. and ZELLWEGER, H. (1965) Blood serotonin deficiency in Down's syndrome. *Lancet*, **2**, 715.

TYCE, G. M., FLOCK, E. V., OWEN, C. A., STOBLE, G. H. and DAVID, C. (1967) 5-Hydroxyindole metabolism in the brain after hepatectomy. *Biochem. Pharmacol.* **16**, 979.

UDENFRIEND, S., WEISSBACH, H. and BRODIE, B. B. (1958) Methods of biochemical analysis, Vol. 6, (ed.) D. Glick, Wiley-Interscience, New York, p. 113.

CHAPTER 2

Administration of 5-Hydroxytryptophan – Open Studies

MARY COLEMAN

Serotonin is an amine in the central nervous system thought to be involved directly or indirectly with neurotransmission. In the blood of many retarded patients, the level of the amine is abnormal for a variety of reasons (Chapter 9). For example, the depressed endogenous level of 5-HT in Down's syndrome reported in the whole blood or platelets may be related to inadequate transport and binding factors in the platelet. If these abnormal levels of 5-HT are also present in the brains of retarded patients, is this contributing to abnormal central nervous system function? Would changing levels of 5-HT in these patients help to understand its role in the central nervous system?

With these considerations in mind, a trial of 5-HT elevation in patients with Down's syndrome was decided upon in 1966. There are a number of ways to do this (see Chapter 8). 5-Hydroxytryptophan (the immediate precursor) was decided upon rather than tryptophan (the natural precursor of 5-HT in the central nervous system) because all of 5-HTP goes directly to 5-HT; only 1% of tryptophan goes into the 5-HT pathway. Also, we had dosage and side effect experience with an amino acid comparable to 5-HTP (L-Dopa) in earlier studies of other disease entities. It is surprising and interesting to note there appears to be some preliminary data (Meier, 1970; Beardsley and Puletti, 1971) showing a return to better intellectual function in Parkinson patients (average age 60 years) taking L-Dopa.

Early studies had shown that 5-HTP raised endogenous 5-HT levels in the central nervous system (Udenfriend *et al.*, 1957; Costa and Aprison, 1958; Cronheim and Gourzis, 1960; Schanberg and Giarman, 1960; Schanberg, 1963). The enzyme that catalyses the decarboxylation of many amino acids is the one used by 5-HTP to make 5-HT. 5-HTP has an advantage over the other substances because it has the lowest Km, that is, the greatest affinity for the enzyme. There have been reports of therapeutic efficacy of tryptophan and concomitant failure of 5-HTP in depressed adults (Carpenter, 1971). Overwhelming peripheral decarboxylation which prevents 5-HTP from reaching the brain has been suggested as the reason for this result. The data in this monograph (Chapters 6 and 7) demonstrate that CNS effects are present in the relatively low dosages of 5-HTP used. They do not prove, however, that the CNS effects recorded are due to increased 5-HT in the serotonergic neurons. They could also be due to mechanisms such as blocking of binding sites of other amines or blocking of active transport systems of other aromatic amino acids, particularly tryptophan itself, the natural CNS precursor of 5-HT. Also, 5-HTP in massive dosages (50 to 500 times the levels in this monograph) may be converted to 5-HT both intraneuronally as previously reported and in many extraneuronal CNS locations not normally containing significant amounts of 5-HT (Corrodi *et al.*, 1967; Fuxe *et al.*, 1971; Bedard *et al.*, 1971).

Our program of administration of 5-HTP to Down's syndrome infants initially was to openly administer the amino acid to study efficacy, method of administration and side effects. Following this, a double blind study of 5-HTP and placebo was planned. This chapter deals with the groups of patients who received 5-HTP alone on an open basis. The next four chapters report on the double blind studies.

In this chapter we are reporting on 14 trisomies who received 5-HTP from the neonatal period. In addition, less detailed studies were done on 12 patients started on 5-HTP at later ages between 3 and 28 months of age. We also gave 5-HTP to a mosaic patient. As an extra study, a group of 22 trisomies received 5-HTP with small doses of vitamin B6.

An elevation of both blood 5-HI and urinary 5-HIAA levels could be demonstrated in most of our patients receiving 5-HTP in low doses and in all our patients receiving more than 5.0 mg/kg. In a paper describing 0.1 to 0.3 mg/kg of 5-HTP infused into terminal liver coma patients, Borges *et al.* (1959) state that 5-HTP raises urinary 5-HIAA but not blood 5-HT at these levels. However, he conducted the 5-HIAA experiment for 24 hours while he limited blood 5-HT data to 3 hours. Davidson *et al.* (1957) also reported no effect on blood 5-HT at these very low levels.

In one 4-month-old patient temporarily overtreated with 5-HTP, the blood 5-HI levels paralleled the mg/kg dosage of the amino acid (Fig. 2-1). Whole blood 5-HI is 99% 5-HT and 1% 5-HIAA in untreated patients and patients receiving less than 0.5 mg/kg of 5-HTP. However, blood total 5-hydroxyindoles in patients receiving 5-HTP above 0.6 mg/kg may have three separate components: 5-HT (57% to 99%), 5-HIAA (1% to 41%) and 5-HTP (0% to 13%). In this overtreated patient, on a dose of almost 100 mg/kg, it was very interesting to observe this Down's syndrome patient's ability to make 1,150 ng/ml of 5-HT; that is, 6–11 times the amount of 5-HT in whole blood of normal children in our laboratory. Her total blood 5-HI was 1,750 ng/ml, with 65% of the total as 5-HT itself (Bazelon *et al.*, 1968).

Early neurological results with Group A (the first group of 14 patients receiving 5-HTP) were described in 1967 (Bazelon *et al.*, 1967). We noted improvements in the Landau posture, traction response, reflexes and activity levels in infant patients after they began receiving 5-HTP by mouth (see Fig. 2-2). In contrast, the adductor component of the Moro was not restored by the amino acid even if the patients were overtreated to the point of hypertonus. We stated "no prognostic inferences regarding intelligence are warranted".

The elevation of platelet serotonin levels was closely correlated with improvement in the neurological examination in the newborns. However, it must be remembered that a good neurological examination at that age is not an indicator of higher cortical functions. The occasionally superior reflex and

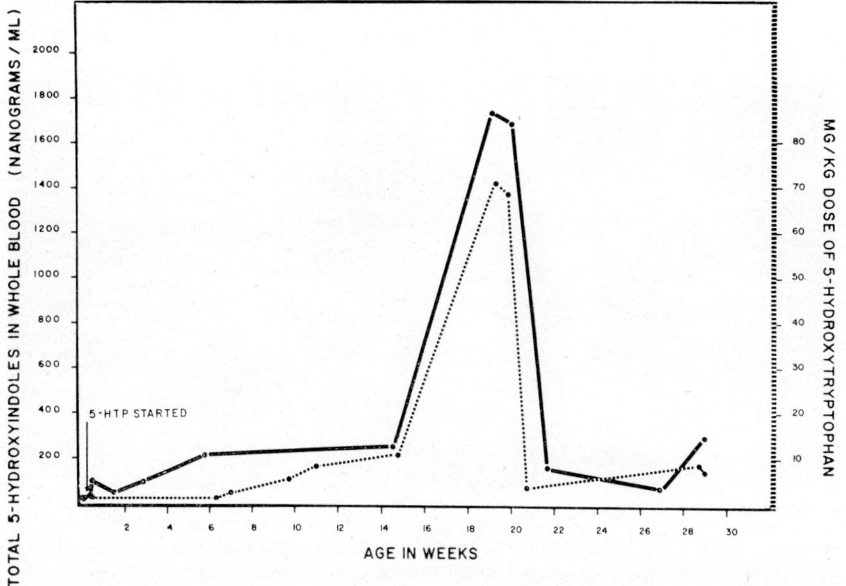

Fig. 2-1. Four-month-old patient A-12 temporarily was given a dose of D,L-5-HTP up to 98.7 mg/kg (dotted line). The pattern of total 5-hydroxyindole values in whole blood (solid line) paralleled the dosage of 5-HTP in this patient with trisomy 21. Sixty-five percent of the total 5-hydroxyindoles were 5-hydroxytryptamine.

infant automatism examination of the hydranencephalic infant in the newborn period is a devastating example of this principle.

In the initial group of 14 patients, dosage levels were higher than those used in the 22 patients who received 5-HTP and B6. This, or the modifying effect of vitamin B6 (seen also in the L-Dopa/B6 therapy), may have accounted for the less striking effect on muscular tonus noted in the group receiving the double research therapy. The mosaic patient was indistinguishable from the trisomies regarding the effect of 5-HTP. No definite effect on tonus could be documented on the 12 patients started on 5-HTP after 6 months of age, although there were questionable activity and alertness effects.

The percentage of Down's syndrome patients with hypotonia ranges from 66% (Levinson *et al.*, 1955) to 98% (McIntire *et al.*, 1964). However, without exact age comparisons, the figures in each series are not really comparable to each other. Many authors have noted that abnormalities in muscular tonus are most striking in Down's syndrome in the younger patients. There even may be X-ray evidence of fetal hypotonicity (Birnbaum, 1971). Like many clinical signs in this disease entity, they improve with age. Øster reports "hypotonia is most pronounced in young children" while hyperflexibility "was found to decrease in frequency with advancing age" (Øster, 1953). Hall in his smaller

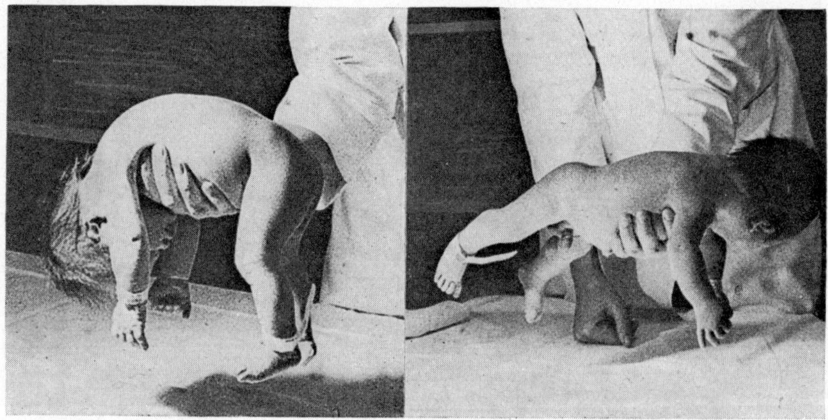

Fig. 2-2. Best Landau posture achieved by patient A-10 at 3 days of age (on left) just before administration of D,L-5-HTP and 24 hours later (on right). On both examinations, time of day, relation to feeding, sleep and previous handling were the same. (Reprinted by permission of the *Lancet*.)

series noted a decrease from "obvious hypotonia" in 35% of newborns to "moderate hypotonia" in 23% of 6-year-olds (Hall, 1970). Penrose reports "with increasing age hypotonia becomes less pronounced" (Penrose and Smith, 1966). However, since we were obtaining improved muscular tonus within a day of administration of 5-HTP, age was ruled out as the critical variable.

Both 5-HTP and the related amino acid, L-Dopa (used in Parkinson's disease (Cotzias, 1968) and dystonia syndromes (Coleman, 1970)), appear to affect muscular tonus. In an interesting study of alpha- and gamma-fiber activity in electromyograms of calf muscle, Roos and Steg (1964) compared the effects of these amino acids to the effects of reserpine. They demonstrated that reserpine decreased gamma-fiber activity and increased alpha-fiber activity, while both L-Dopa and 5-HTP reversed this effect. Their conclusion was that the rigidity (increased tonus) in Parkinson's disease was similar to reserpine-induced increased tonus.

How 5-HT may affect tone is unknown because the exact biochemical mechanism of action of 5-HT in the central nervous system is unknown. One of the most interesting recent suggestions is that 5-HT may exert its effect at specific receptor sites by influencing the concentration of potassium, and possibly chloride ions (Hanig and Aprison, 1971). Any compound regulating the transport of potassium across postsynaptic or other neuronal membranes will affect the resting potential, thus modulating the excitatory and inhibitory transmissions within the central nervous system. This study also compared 5-HTP to reserpine; however, it showed that both compounds raised potassium levels.

Pryor and Mitva (1970) severely depleted rats of 5-HT from birth to 6 weeks by giving them *p*-chlorophenylalanine (*p*CPA), a tryptophan hydroxylase inhibitor. Their 'ambulation' was significantly and markedly depressed while receiving *p*CPA but both motor function and maze-solving abilities came back to normal after *p*CPA was stopped. The authors considered this compelling evidence that reduction in brain serotonin in the postnatal development does not lead to mental retardation. (They may be right but their experiment doesn't prove it since among various difficulties with the interpretation is the fact that the developmental period in a rat brain and human brain do not match at the time of birth (W. Himwich, personal communication).) In any event, the temporary decrease in ambulation when serotonin was lowered is of interest, although just feeling toxic from a large dose of any drug can affect activity.

The changes in muscular tonus noted on the 5-HTP infants led to the first question we decided to investigate in the double blind study

Postulate One

Are the changes in tonus observed in patients receiving 5-HTP due to the amino acid? If so, how many months do they persist on chronic 5-HTP administration?

Muscle tonus is one factor that determines the age at which independent walking begins. Another factor is the extrapyramidal control of balance, by complex interaction between vestibular, brainstem and cerebellar systems. Cerebellar function is thought to be disturbed in Down's syndrome (Cowie, 1967). However, as the Group A patients matured past 1 year of age, the striking truncal ataxia and loss of balance described in the literature was not obvious. It was surprising to watch the first patient in the study bend over without falling and pick a marble off the floor at 18 months of age.

The brain is usually small in Down's syndrome and the brainstem and cerebellum are the sections most decreased in size (Wilmarth, 1890; Crome and Stern, 1967; Rorke *et al.*, 1968). In 1966, Crome *et al.* suggested that there may be a causal connection between the decreased size of the cerebellum and brain-stem and the profound hypotonia in Down's syndrome. When the 5-HTP project was in its second year and noted apparent improvement in balance, it was difficult to attribute it to improved cerebellar function. The literature clearly states that in animals (Garattini and Valzelli, 1958) and adult humans (Costa and Aprison, 1958) there is relatively little 5-HT in the cerebellum. However, we have recently examined the 5-HT content of the brains of two young children and found that in both those brains the central cerebellar nuclei have significant amounts of 5-HT (Chapter 9).

It is of interest that the 7 5-HTP patients taken off the amino acid very rapidly instead of being tapered off experienced loss of balance (with loss of ability to pull to standing in the youngest child) as the one common finding (Appendix II-1). One of these patients (D-9) had also experienced ataxia as a sign of overtreatment. In some cases, elevation of 5-HI levels by increased 5-HTP (patients A-9 and B-10) or reinstitution of B6 alone (patient D-9) reversed the ataxia. Patients C-14 and C-31 remained off the 5-HTP and stayed ataxic for months; they slowly overcame it as they progressed developmentally.

Whenever an amino acid is abruptly withdrawn, there is always the problem of interpreting the immediate effects. Are they due to loss of clinical effect of the amino acid or are they acute withdrawal symptoms? This problem also exists in interpreting crossover studies with L-Dopa (Coleman, 1970). (With amino acids, an extra Koch's postulate needs to be written—"the amino acid may be presumed to have suppressed symptoms only if these symptoms return as the amino acid is tapered down slowly rather than withdrawn abruptly".)

A later evaluation of tonus and equilibrium in our patients was based on the age of walking. The home-reared Group A patients walked, on the average, at 22 months (see Appendix II-2). These are the patients reported in our paper in *Lancet* in 1967 (Bazelon *et al.*, 1967). This is earlier than the other home-reared series in the literature (Table 2-1). In view of the importance of home rearing versus institutionalization in establishing age of walking (Centerwall and Centerwall, 1960; Donaghue *et al.*, 1967), the Brousseau (1928) series averaging 36 months as the age of walking was not included since it combines both home-reared and institutionalized patients. Paulson (1971) studied institutionalized patients and he believes "major psychologic factors may be contributory . . . minor alterations in the environment, such as a

TABLE 2-1
Age of walking in Down's syndrome

	No. of patients	Mean age of walking (months)	References
Home-reared until age of walking			
Group A	13	22	Appendix II-2
Fishler *et al.* (1964)	50	25	Fishler *et al.* (1964)
Levinson *et al.* (1955)	50	26	Levinson *et al.* (1955)
Erbs and Smith (1967)	—	27	Donaghue *et al.* (1967)
Donaghue *et al.* (1967)	42	28	Donaghue *et al.* (1967)
Centerwall and Centerwall (1960)	28	30	Centerwall and Centerwall (1960)
Institutionalized before age of walking			
Group A (A-14)	1	35	Appendix II-2
Donaghue *et al.* (1967)	20	53	Donaghue *et al.* (1967)
Centerwall and Centerwall (1960)	17	49	Centerwall and Centerwall (1960)

change in attendants or cessation of parental visits may be the main cause for reduced ambulation . . ." The home-reared Group A patient (A-6) who walked the latest (40 months) was the only child in the group who did not receive dosages as prescribed (based on pharmacy calculations). Also, she was raised primarily by servants because of rejection by her mother, who made three suicide attempts during the first year of the patient's life and was said to have even run away from home on some weekends when the task of taking care of the child was entirely her responsibility. The one patient in Group A (patient A-14) who was raised in an institution also walked late, at 35 months. These results support other data showing the importance of environmental factors and parental attitudes in determining the age of walking. One could make a supposition that 5-HTP (if it was administered as prescribed and if it had an effect on walking) did little, if anything, to overcome the overwhelming effects of environment in patients A-14 and A-6.

Conversely, in evaluating the earlier onset of walking in most Group A patients, it is important to keep in mind the selected type of patients with good home environments in our clinic (Appendix I-2b).

This led to the second question of the double blind study:

POSTULATE TWO

Are 5-HTP or placebo, environmental, chance or selection factors responsible for the earlier onset of the age of walking noted in Down's syndrome patients receiving the amino acid compared to other series in the literature?

Several parents reported that raising or lowering the dose of 5-HTP affected the strabismus present in their children. Since fluctuation of strabismus often occurs in Down's syndrome patients (Lowe, 1949; Cowie, 1970), it was difficult to be sure if these reports had any validity. However, several authors (Engler, 1949; Cowie, 1970) report that at least one component of strabismus in this patient group is related to hypotonia of the ocular muscles. Cowie, in her detailed computer study, found a significant positive correlation between strabismus and marked hypotonia in her patients as well as other evidence that strabismus was an independent variable.

This led to our third question of the double blind study:

Postulate Three

Do patients receiving 5-HTP have any difference in frequency or pattern of strabismus compared to patients on placebo?

Another unusual thing about the Group A patients was the total absence of cardiac disease in the series. Very early in the initial study, two patients had been declared ineligible for the study—one, because she developed cardiac symptoms within 48 hours of birth and a second child, because he had no elicitable EEG auditory evoked potentials and could not be followed by this research technique. There was other evidence of selective factors in the Group A patients, i.e., in this initial study, 2 were children of physicians. However, in spite of the bias and the fact that 1 patient had been turned down because of cardiac disease, we found it surprising that not a single remaining one of the 14 patients in Group A had any cardiac disease (Appendix II-3). In Group D (the group of patients receiving 5-HTP and B6), a similar trend was seen. Only 5% (1 out of 22) had cardiac symptoms, although no patient was omitted from that study for any medical reason.

The exact incidence of cardiac disease in Down's syndrome is not established, although it has been almost 80 years since Garrod first noted the increased frequency in these patients (Garrod, 1894). Bertil Hall examined every newborn with Down's syndrome born during a 12-month period in the four southern counties of Sweden (Hall, 1964). By studying babies immediately after birth, he even was able to include a child who died by 10 hours of age; Hall has the most complete prospective survey in the literature. Review of his group when the patients were 6 years of age showed that 14 out of his 38 patients had cardiac disease (37%); in 10 of these, the diagnosis was confirmed by autopsy (Hall, 1970). Only 13% of the patient group still alive at 6 years of age had cardiac symptoms.

Another Scandinavian, Øster, did an extensive retrospective survey of Down's syndrome in three counties of Denmark covering a 26-year period from 1923 to 1949 (Øster, 1953). He felt his survey was incomplete because "it is practically impossible to trace all cases". However, it is more complete than any other large series. In his series, cardiac auscultation disclosed cardiac disease in 12.4% (60 out of 423) patients living at the time of the survey.

Thus, in the 2 studies based as closely as possible on geographical ascertainment, the percentage of surviving patients with cardiac disease is similar—between 12 and 13% after infancy.

Other series of patients (based on autopsy studies, cardiac clinics or institutions) suffer from major factors of bias. Autopsy studies are biased toward cardiac patients since it is the major cause of death in young patients (Øster, 1953; Berg *et al.*, 1960; Evans, 1950; Esen, 1957) and these series show high percentages—up to 75% (Benda, 1969) 56% (Berg *et al.*, 1960). Studies from cardiac clinics suffer from bias related to admission procedures of the clinic. In spite of the limitation, Rowe and Uchida (1961) did a careful study and demonstrated cardiac anomalies in 40% of a large series. Institutional surveys, such as the Øster (1953) report of 26% cardiac patients among all Down's syndrome admissions to pediatric hospital wards, favor the ill patient with congestive cardiac disease.

There have been studies which suggest that 5-HT causes a selective increase in pulmonary vascular resistance compared to the systemic circulation. Such an increase in pulmonary vascular resistance theoretically could help compensate for several of the major cardiac defects in Down's syndrome, and possibly could account for decreased cardiac symptoms.

The low incidence of such symptoms in 5-HTP patients led to our fourth question.

Postulate Four

Are 5-HTP, placebo, environmental effects, selection bias or chance responsible for the low incidence of cardiac symptoms in Down's syndrome patients receiving the amino acid in the initial 5-HTP study?

The size and type of tongue protrusion in Down's syndrome is a controversial subject. Some clinicians writing on the subject believe that the tongue is macroglossic: "large in 57%" (Øster, 1953) "an abnormally large tongue" (Poser, 1969). Other authors attribute the tongue protrusion to "shortness of the oral cavity causing protrusion of the tongue" (Brousseau, 1928), "smallness of oral cavity and hypoplasia of the mandible" (Nelson, 1969). Shapiro *et al.* (1967) have shown that the length of the palate is shorter and that the width is also decreased in this patient group. Penrose and Smith (1966) report "many authors, however, consider that the tongue is normal in size and only in rare cases it is enlarged".

Our observations suggest that although somewhat decreased in size, the oral cavity is large enough to contain the usually normal sized tongue and that the protrusion is, in fact, of central nervous system origin. There appear to be two types of tongue visibility or protrusion in Down's syndrome patients: (1) buccal–lingual hypotonia—laxness, thickness, relaxation of the tongue which is visible between slack, open lips, usually accompanied by dullness of expression in younger patients or an expression of focused concentration in older patients (Fig. 2-3), and (2) buccal–lingual dyskinesia—a darting, intermittent protrusion of the tongue in and out of the lips: sometimes pointing (snaking) or curling of the tongue is seen and noises (raspberrying) accompanying the motion (Figure 2-4). Both types of tongue protrusion can be seen in Down's syndrome patients not receiving 5-HTP; primarily the buccal–lingual

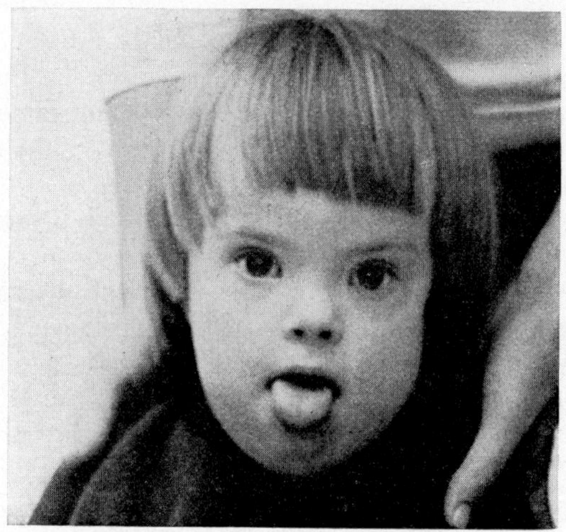

Fig. 2-3. Buccal–lingual hypotonia.

dyskinesia is seen in patients receiving toxic doses of the amino acid. The buccal–lingual hypotonias and dyskinesias are seen in many of the serotonin syndromes (see Chapter 9).

Both types of tongue protrusion can also be seen in normal children particularly when they are tired, bored, hungry, teething or with nasopharyngeal stuffiness. However, even as small babies, normal children rarely have the amount of tongue protrusion seen in most Down's syndrome patients.

In Down's syndrome the natural course of the tongue protrusion is similar to hypotonia; in most cases, it eventually disappears with time although at a considerably later age than in those normal children who also happen to have it. Øster reports (what he calls) macroglossia is "most frequent in young age groups" (1953). Hall (1970) reports only 9% of his series had a protruding tongue at 6 years of age and listed it in his classification of signs that diminish or disappear.

Clinical observation of patients placed on 5-HTP suggested that buccal–lingual dyskinesia was increased in some patients by 5-HTP. Buccal–lingual dyskinesias of a more bizarre type have previously been induced in adult patient groups who receive overdosage of L-Dopa (Barbeau and McDowell, 1970) or phenothiazines (Shepherd *et al.*, 1968). Both these compounds affect 5-HT (Ritvo *et al.*, 1971; Tyce *et al.*, 1970; Everett and Borcherding, 1970; Goodwin *et al.*, 1971). However, the relationship is a complex one since adding 5-HTP to the regimen of an advanced Parkinson patient who had

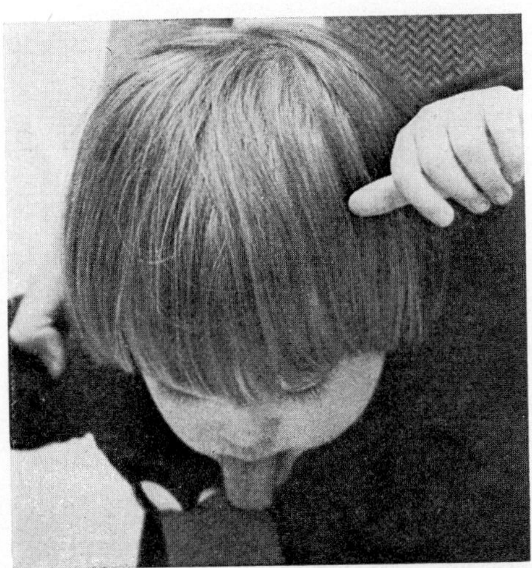

FIG. 2-4. A Down's syndrome patient with buccal–lingual dyskinesia has a darting, intermittent protrusion of the tongue, sometimes quite far out of the oral cavity.

L-Dopa induced buccal–lingual dyskinesia made the dyskinesia worse (Fig. 2-5).

Also, the presence of buccal–lingual hypotonia or dyskinesia does not necessarily mean that there is a serotonin abnormality in the platelet; we have measured normal 5-HI levels in several retarded patients with tongue protrusion. Conversely, a small portion of Down's syndrome patients have no tongue protrusion, despite low 5-HI levels in the blood.

POSTULATE FIVE

Is 5-HTP a factor in the changed pattern of tongue visibility reported in the patients receiving the amino acid?

There is evidence in the literature that 5-HT may be related to temperature regulation. We decided to do a spot check of morning temperatures of the patients during their initial admission to the Clinical Research Center for their evaluations.

Postulate Six

Does administration of 5-HTP affect baseline temperatures in Down's syndrome infants?

Many observers thought an increase in alertness and activity level was noted in the original groups of patients when they received 5-HTP. Occasionally patients appeared restless and too active. Some patients turned over

Length of buccal-lingual dyskinesia in Parkinson patients on L-Dopa

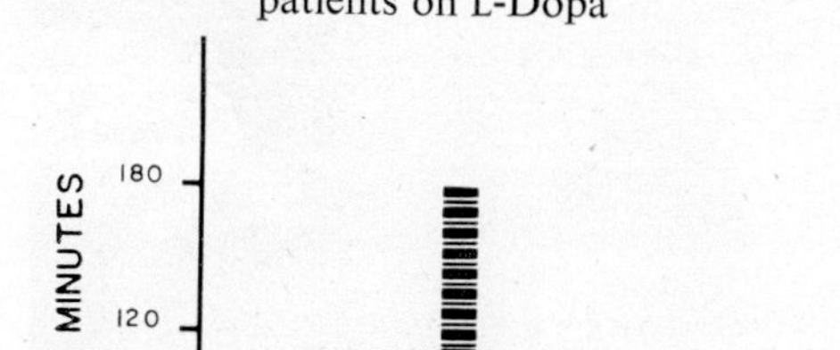

Fig. 2-5. A 68-year-old male, with a 16-year course of idiopathic Parkinson's disease, was barely able to tolerate 10 mg/kg of L-Dopa daily because of the side effect of buccal–lingual dyskinesia. The addition of 2.5 mg/kg of L-5-HTP lengthened the time the dyskinesia persisted after the dose of both amino acids.

unusually early (first to third week) from supine to prone and one frantic 12-day-old turned over both ways (recorded on motion picture film). Since increased activity can be accompanied by increased catecholamine turnover, a measurement of catecholamine metabolites in the urine appeared to be an objective way of comparing the two patient groups.

We also had planned to look at catecholamines for another reason. In the many studies of the effect of large doses of L-Dopa upon 5-HT metabolites (Ritvo *et al.*, 1971; Goodwin *et al.*, 1971; Van Woert and Bowers, 1970; Everett and Borcherding, 1970; Tyce *et al.*, 1970) a consistent decrease of these metabolites has been reported. We had seen the same effect on whole blood 5-HI levels when we administered L-Dopa to patients with elevated whole blood 5-HI levels, as seen in Fig. 2-6. Would 5-HTP have a comparable effect of lowering catecholamine levels in patients receiving large doses of 5-HTP? Since we are unaware of a comparable 'model' system for catecholamines as exists with the platelet 5-HT model system, it was decided to rely on the 24-hour excretion of catecholamines and their metabolites.

POSTULATE SEVEN

Is there any objective evidence of increased activity in patients receiving 5-HTP, such as an increase in epinephrine in the urine? How are the catecholamines and their metabolites affected by large doses of 5-HTP?

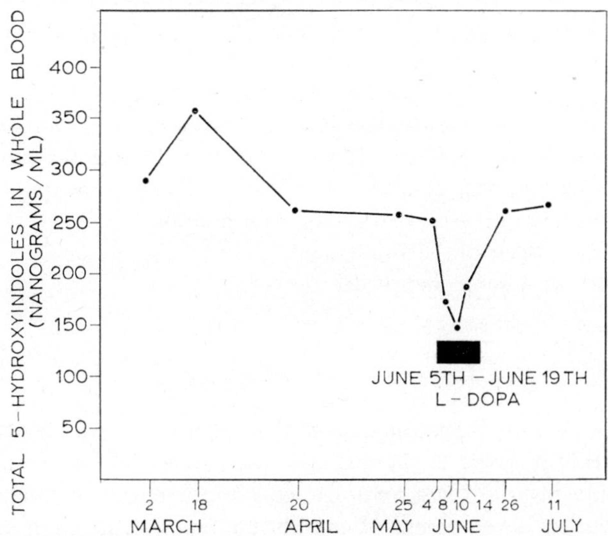

Fig. 2-6. A 2-year-old patient with the infantile spasms syndrome had a consistently elevated endogenous level of whole blood 5-HI except during the period she received 25 mg/kg of L-Dopa.

Occasionally very active patients in the early studies failed to gain weight well and two patients lost weight coincidentally during a period when they were on high doses of 5-HTP. Some recent studies have shown that administration of amino acids such as phenylalanine and L-Dopa to young rats can cause disaggregation of brain polysomes and inhibition of brain synthesis (Oaki and Siegel, 1970; Weiss *et al.*, 1971). Studies with 5-HTP have not been performed. The doses used were much higher than any described in this monograph. Essman (1970) also demonstrated in animals that high elevation of brain 5-HT may be related to decreased RNA and decreased incorporation of an amino acid into protein. In a patient group already retarded and short in stature, administration of amino acids needs careful evaluation.

Postulate Eight

Is there any difference in the height, weight and head sizes of patients chronically receiving 5-HTP and patients on placebo?

And finally, perhaps most important of all, a comparison of 5-HT levels in blood and urine in the patients was needed.

Postulate Nine

Is the whole blood 5-HI and urinary 5-HIAA significantly elevated in patients receiving 5-HTP compared to placebo patients? What is the relationship of the placebo 5-HI levels to the first visit baseline 5-HI levels of our clinic?

These postulates are discussed in Chapter 3. Additional questions asked by a psychologist, psychiatrist and EEG specialist regarding the double blind study are in the following chapters. The trial of 5-HTP in the initial group of patients, including one toxic overdosage at 4 months of age, did not indicate any serious side effects of the amino acid.

It was time for a controlled study.

Summary

This chapter presents the findings in the initial group of patients (Group A) receiving 5-HTP that led to the decision to conduct a double blind study of 5-HTP administration. These findings were improvement of muscle tone in newborns, earlier development of equilibrium and balance, improvement in the mean age of walking, increased activity levels, parental reports of improvement in strabismus and tongue protrusion with variation of 5-HTP doses, and lack of cardiac disease in a (selected) initial group of patients. The series of postulates to be studied in a 5-HTP/placebo double blind study were formulated from these findings. The double blind study is described in the next chapter.

REFERENCES

Barbeau, A. and McDowell, F. H. (1970) L-Dopa and parkinsonism, F.A. Davis Co., Philadelphia, Pa.

Bazelon, M., Barnet, A., Lodge, A. and Shelburne, S. (1968) The effect of high doses of 5-hydroxytryptophan on a patient with trisomy 21, clinical chemical and EEG correlations. *Brain Research*, **11**, 397.

Bazelon, M., Paine, R. S., Cowie, V., Hunt, P., Houck, J. C. and Mahanand, D. (1967) Reversal of hypotonia in infants with Down's syndrome by administration of 5-hydroxytryptophan. *Lancet*, **1**, 1130.

BEARDSLEY, J. V. and PULETTI, F. (1971) Personality (MMP1) and cognitive (WAIS) changes after levodopa treatment. *Arch. Neurol.* **25**, 145.

BEDARD, P., CARLSSON, A., FUXE, K. and LINDQVIST, M. (1971) Origin of 5-hydroxytryptophan and L-Dopa accumulating in brain following decarboxylase inhibition. *Naunyn-Schmiedeberg's Arch. Pharmak.* **269**, 1.

BENDA, C. E. (1969) Down's syndrome–mongolism and its management, Grune and Stratton, New York.

BERG, J. M., CROME, L. and FRANCE, N. E. (1960) Congenital cardiac malformations in mongolism. *Brit. Heart J.* **22**, 331.

BIRNBAUM, S. J. (1971) Prenatal diagnosis of mongolism by X-ray. *Obstet. and Gynec.* **37**, 394.

BORGES, F. J., MERLIS, J. K. and BESSMAN, S. P. (1959) Serotonin metabolism in liver disease. *J. Clin. Invest.* **38**, 715.

BROUSSEAU, K. (1928) Mongolism, Williams and Wilkins, Baltimore.

CARPENTER, W. T. J. (1971) Serotonin in affective disorders. *Ann. Int. Med.* **73**, 613.

CENTERWALL, S. A. and CENTERWALL, W. R. (1960) A study of children with mongolism reared in the home compared to those reared away from the home. *Pediatrics*, **25**, 678.

COLEMAN, M. (1970) Preliminary remarks on the L-Dopa therapy of dystonia. *Neurology*, **20**, 114.

CORRODI, H., FUXE, K. and HOKFELT, T. (1967) Replenishment by 5-hydroxytryptophan of the amino stores in the central 5-hydroxytryptamine neurons after depletion induced by reserpine or by an inhibition of monoamine synthesis. *J. Pharm. Pharmacol.* **19**, 433.

COSTA, E. and APRISON, M. H. (1958) Studies on the 5-hydroxytryptamine (serotonin) content in human brain. *J. Nerv. Ment. Dis.* **126**, 289.

COTZIAS, G. C. (1968) L-Dopa for parkinsonism. *New Engl. J. Med.* **278**, 630.

COWIE, V. A. (1967) Neurological aspects of the early development of mongols. *Clin. Proc. Children's Hosp.* (*Washington, D.C.*), **23**, 64.

COWIE, V. A. (1970) A Study on the Early Development of Mongols, Pergamon Press, Oxford.

CROME, L., COWIE, V. and SLATER, E. (1966) A statistical note on cerebellar and brain-stem weight in mongolism. *J. Ment. Defic. Res.* **10**, 69.

CROME, L. and STERN, J. (1967) The pathology of mental retardation, Ly. and A. Churchill Ltd., London.

CRONHEIM, G. E. and GOURZIS, J. T. (1960) Cardiovascular and behavioral effects of serotonin and related substances in dogs without and with reserpine premedication. *J. Pharmacol. Exp. Ther.* **130**, 444.

DAVIDSON, J., SJOERDSMA, A., LOOMIS, L. N. and UDENFRIEND, S. (1957) Studies with the serotonin precursor, 5-hydroxytryptophan, in experimental animals and man. *J. Clin. Invest.* **36**, 1594.

DONAGHUE, E. C., KIRMAN, B. H., LABAN, D. and ABBAS, A. (1967) Age of walking in the mentally retarded, in: Proceedings, First Congress of the International Association for the Scientific Study of Mental Retardation, Montpellier.

ENGLER, N. (1949) Mongolism (peristatic amentia), Williams and Wilkins, Baltimore.

ERBS and SMITH (1967) in: Donaghue, 1967.

ESEN, F. M. (1957) Congenital heart malformation in mongolism with special references to ostium atrioventricular commune. *Arch. Pediat.* **74**, 243.

ESSMAN, W. B. (1970) Some neurochemical correlates of altered memory consolidation. *Trans. N.Y. Acad. Sci.* **32**, 948.

EVANS, P. R. (1950) Cardiac anomalies in mongolism. *Brit. Heart J.* **12**, 258.

EVERETT, G. M. and BORCHERDING, J. W. (1970) L-Dopa: effect on concentrations of dopamine, norepinephrine, serotonin in brains of mice. *Science*, **168**, 849.

FISHLER, K., SHARE, J. and KOCH, R. (1964) Adaptation of Gesell developmental scales for evaluation of development in children with Down's syndrome (mongolism). *Am. J. Ment. Def.* **68**, 642.

FUXE, K., BUTCHER, L. L. and ENGEL, J. (1971) D,L-5-Hydroxytryptophan-induced changes in central monoamine neurons after peripheral decarboxylase inhibition. *J. Pharm. Pharmac.* **23**, 420.

GARATTINI, S. and VALZELLI, L. (1958) Researches on the mechanism of reserpine sedative action. *Science*, **128**, 1278.

GARATTINI, S. and VALZELLI, L. (1965) Serotonin, Amsterdam, Elsevier.

GARROD, A. E. (1894) On the association of cardiac malformation with other congenital defects. *St. Barth. Hosp. Rep.* **30**, 53.

GOODWIN, F. K., DUNNER, D. L. and GERSHON, E. S. (1971) Effect of L-Dopa treatment on brain serotonin metabolism in depressed patients. *Life Sci.* **10**, 751.

HALL, B. (1964) Mongolism in newborns. *Acta Paediat.* (*supplement*), **154**, 3.

HALL, B. (1970) Somatic deviations in newborn and older mongoloid children. *Acta Paediat. Scand.* **59**, 199.

HANIG, R. C. and APRISON, M. H. (1971) The effect of 5-hydroxytryptophan and reserpine administration on the level of sodium, potassium, calcium, magnesium and chloride in five discrete areas of the rabbit brain. *Life Sci.* **10**, 279.

LEVINSON, A., FRIEDMAN, A. and STAMPS, F. (1955) Variability of mongolism. *Pediatrics*, **16**, 43.

LOWE, R. F. (1949) The eyes in mongolism. *Brit. J. Ophthalmol.* **33**, 131.

MCINTIRE, M. S., MENOLASCINO, F. J. and WILEY, J. H. (1964-65) Mongolism–some clinical aspects. *Am. J. Ment. Def.* **69**, 794.

MEIER, M. J. (1970) Intellectual changes associated with levodopa therapy (letter). *J. Am. med. Ass.* **213**, 465.

NELSON, W. E. (1969) Textbook of pediatrics, W. B. Saunders Company, Philadelphia.

OAKI, K. and SIEGEL, F. L. (1971) Hyperphenylalanemia; disaggregation of brain polyribosomes in young rats. *Science*, **168**, 129.

ØSTER, J. (1953) Mongolism. A clinicogenealogical investigation comprising 526 mongols living on Seeland and neighbouring islands in Denmark, Danish Science Press, Ltd., Copenhagen.

PAULSON, G. W. (1971) Failure of ambulation in Down's syndrome; a clinical survey. *Clin. Pediat.* (*Phila.*), **10**, 265.

PENROSE, L. S. and SMITH, G. F. (1966) Down's anomaly, J. and A. Churchill Ltd., London.

POSER, C. M. (1969) Mental retardation, diagnosis and treatment, Harper and Row, Hoeber Medical Division, New York.

PRYOR, G. T. and MITVA, C. (1970) Use of p-chlorophenylalanine to induce a phenylketonuric-like condition in rats. *Neuropharmacology*, **9**, 269.

RITVO, E. R., YUWILER, A., GELLER, E., KALES, A., RASHKIS, J., SCHICOR, A., PLOTKIN, S., AXELROD, R. and HOWARD, C. (1971) Effects of L-Dopa in autism. *J. Aut. Childhood Schiz.* **1**, 190.

ROOS, B. E. and STEG, G. (1964) The effect of L-3,4-dihydroxyphenylalanine and DL-5-hydroxytryptophan on rigidity and tremor induced by reserpine, chlorpromazine and phenoxybenzamine. *Life Sci.*, **1**, 351.

RORKE, L. B., FOGELSON, M. H. and RIGGS, H. E. (1968) Cerebellar heterotopia in infancy. *Develop. Med. Child Neurol.* **10**, 644.

ROWE, R. D. and UCHIDA, I. A. (1961) Cardiac malformations in mongolism. *Amer. J. Med.* **31**, 726.

SCHANBERG, S. M. (1963) A study of the transport of 5-hydroxytryptophan and 5-hydroxytryptamine (serotonin) into brain. *J. Pharmacol. Exp. Ther.* **139**, 191.

SCHANBERG, S. M. and GIARMAN, N. J. (1960) Uptake of 5-hydroxytryptophan by rat brain. *Biochim. Biophys. Acta*, **41**, 556.

SHAPIRO, B. L., GORLIN, R. J., REDMAN, R. S. and BRUHL, H. H. (1967) The palate and Down's syndrome. *New Engl. J. Med.* **276**, 1460.

SHEPHERD, M., LADER, M. and RODNIGHT, R. (1968) Clinical pharmacology, Lea and Febiger, Philadelphia.

SOLITARE, G. B. (1969) The spinal cord of the mongol. *J. Ment. Def. Res.* **13**, 1.

TYCE, G. M., FLOCK, E. V. and TAYLOR, W. F. (1970) Effect of ethanol on 5-hydroxytryptamine turnover in rat brain. *Proc. Soc. Exp. Biol. Med.* **134**, 40.

UDENFRIEND, S., WEISSBACH, H. and BOGDANSKI, D. F. (1957) Increase in tissue serotonin following administration of its precursor 5-hydroxytryptophan. *J. Biol. Chem.* **224**, 803.

VAN WOERT, M. H. and BOWERS JR, M. B., (1970) The effect of L-Dopa on monoamine metabolites in Parkinson's disease. *Experientia*, **26**, 161.

WEISS, B. F., MUNRO, H. N. and WURTMAN, R. J. (1971) L-Dopa: disaggregation of brain polysomes and elevation of brain tryptophan. *Science*, **173**, 833.

WILMARTH, S. W. (1890) Report on the examination of one hundred brains of feeble minded children. *Alienist and Neurologist*, **11**, 520.

CHAPTER 3

A Double Blind Trial of 5-Hydroxytryptophan in Trisomy 21 Patients

MARY COLEMAN and LOUIS STEINBERG

INTRODUCTION

After the results noted on muscular tonus by administration of 5-HTP in the initial group of infant patients, detailed further study was indicated. A double blind study of administration of the amino acid to patients with the trisomy 21 form of Down's syndrome was chosen.

We chose this form of study well aware of the limitations of double blind studies for our project. Since the dosage of 5-HTP and placebo was individualized in each patient, the probability of side effects appearing as the dose was raised (5-HTP), or *not* appearing at very high dosages (placebo) would tend to alert the examiner. We considered the possibility of using an 'active' placebo, one that would have similar side effects, but could not decide on one unequivocably safe for administration to the immature central nervous system over a 3-year period.

Another problem was the importance of proper dosages of the amino acid to give an adequate trial of efficacy. The initial double blind studies of the effect of L-Dopa (a cousin amino acid of 5-HTP) in Parkinson's disease had shown no difference from placebo (Fehling, 1966; Rinne and Sonninen, 1968). These studies were inadequate since it is now believed that L-Dopa treatment of Parkinson's disease is the "most important contribution to medical therapy of a neurological disease in the past 50 years" (*New Engl. J. Med.* Editorial, 1971). Recently, well after everyone in the field knew that it worked, the L-Dopa treatment of Parkinson's has been 'blessed' by a positive double blind study (Muenter, 1970). Like any other study, a double blind study is only as good as the techniques used in performing it.

A third problem related to completing the 5-hydroxyindole studies in blood and urine but not using them to help determine dosage levels as we had with the first group of patients. When setting up the protocol, we believed that seeing the 5-HI levels would definitely reveal which patient was on treatment. Therefore, no 5-HI levels at all were seen by the examiners for the first

12 months of the study; they were completed in the laboratory and locked away. In the first group of patients, the hypotonia examination became progressively unrelated to 5-HI levels after 6–12 months of age. So we set up the study with access to the 5-HI levels, if needed to determine dosage, after 1 year of age. As detailed in the biochemical section of this chapter, the 5-HI levels did not turn out to be as reliable an indication of 5-HTP administration as anticipated. In some of the placebo patients, seeing these levels later in the study tended to confuse rather than enlighten the examiners. We knew that by not using 5-HI levels at all in the first year of life, we limited our ability to titrate dosages finely and decreased the value of the study, but felt it was necessary for the sake of as much complete 'blindness' as possible.

Patient Selection

At the time of initiation of the double blind study, we had had a Down's syndrome clinic already in operation for a year and one-half. All patients referred were accepted immediately into the study without exception and placed on a research solution within 24 hours of admission to the research center. Eight patients who had been accepted were eventually dropped from the study—3 died within 2 months of age, 3 were placed in institutions making them ineligible for this study, 1 was found to be a mosaic and 1 was a translocation (for details see Appendix III-1). Seven of these patients were replaced by new referrals so that the total number of patients in the study group was 19 patients. Because of failure of chromosomal culture growths, the translocation patient was not detected until after the study had been closed and could not be replaced. Counting the 8 patients omitted from the final study group, we admitted 27 patients to the double blind study. The infant death rate in our study (11%) was considerably lower than the 39% in the geographically ascertained Hall study (1970).

Characteristics of the families of the 19 patients in the final study group are seen in Appendix III-2. They reflect the same pattern as seen in our clinic as a whole, with a predominant white middle class population with incomes considerably higher than the national average (see Appendix I-2b). A major difference with other published series was the frequency (32%) of patients who had been exposed to pelvic radiation prior to conception.

Protocol of the Study

Twenty letters of the alphabet were randomly assigned by the pharmacologist to L-5-HTP (5 letters), D,L-5-HTP (5 letters) and placebo (10 letters).

The patients were admitted to the study between 12 hours to 6 days of age. The father (or other close relative) met with the neurologist and discussed the study. After signing permission for the study, he then drew a card marked with one of the 20 randomly assigned solution letters from a vase in the presence of a witness. This became the patient's solution for the study.

Each patient was hospitalized for the first 3 weeks of life in a Clinical Research Center and again for a series of 2 day hospitalizations in the Center at the ages of 1, 2 and 3 years.

Adjustment of 5-HTP dose was based on neurological examination only (year 1) and neurological examination plus whole blood 5-HI as needed (in years 2 and 3). Between hospitalizations, the patient was seen in the Down's syndrome out-patient clinic at regular intervals for dosage adjustments; the visits were most frequent in the first year of life.

We administered more than 5-HTP at our clinic. We did everything we could to counteract the prevailing negative attitude toward these patients. Each parent was given a copy of the book '*The World of Nigel Hunt*: *The Diary of a Mongol Boy*', a story written by a child with trisomy 21. The parents were told that to enjoy and love their children was as important to brain development as giving the 'medicine'. As one of the parents expressed it later in a popular magazine, "These children, it has been found, respond to love and attention like flowers to the sun" (Trainer, 1971). Parents with depressive reactions to their child were referred for intensive, immediate professional help. The parents organized themselves into a group and became actively involved in assisting the clinic. In short, the placebo effect (the sole therapy in half the double blind patients) was maximized in all clinic patients.

Breaking the Double Blind Code

After the 3-year evaluation of each patient, the parents and the examiners opened the envelope containing the double blind code on the patient. Several parents whose children were on placebo refused to believe it, quoting their own experience with 'side effects' of the placebo. One placebo patient had suffered from lethargy, diarrhea and a rash after receiving an accidental overdose of the unknown solution. Another parent, whose child was on placebo, quoted many examples of improvement in tone and tongue protrusion with increasing the dose of the solution and demanded to see the pharmacologist who held the double blind code, insisting there must have been a mistake. The general literature and popular concept of Down's syndrome is so linked to an unattractive, inevitably severely retarded patient that many placebo parents felt their child *must* have been receiving a 'treatment' solution to have been such an attractive baby and young child who was learning (albeit, slowly). The 'treatment' their child received was not from a magical biochemical solution but from his own parents.

Family Situations

Family situations were stable in the majority of cases. One set of parents (B-12) separated and divorced during the study. Patient B-13 was an illegitimate child brought into a home with her self-supporting mother's three other children. In spite of these difficulties, her mother never missed an appointment. Patient B-14's father entered a hospital for a number of months during the second year of the study. Patient B-17's parents almost separated when the child was 20 months of age. The mother became depressed and was considering divorce, but they remained together until the termination of the study. At the 3-year evaluation, the father was noted to be both bored and hostile.

Three patients in the study, B-1, B-3 and B-6 had 5 or more other siblings. It became clear that the investment of the parents in their abnormal child was not very great and this was manifested by a poor relationship between the child and the parents. Patient B-6 was the most accepted by her parents of the three, although she preferred her father over her mother. In the case of patient B-1, at the time of the 3-year examination, the child resisted feeding by the mother and was said to 'hate' her, but was affectionate toward all others. An even more rejecting situation existed initially for patient B-3 discussed later in this chapter.

Postulate One

Are the changes in tonus observed in patients receiving 5-HTP due to the amino acid? If so, how many months do they persist on chronic 5-HTP administration?

Regarding muscle tonus, the patients destined to receive 5-HTP started out with a disadvantage; they had a considerably lower tonus score than the patients destined to receive placebo (see Appendix III-3 and Fig. 3-1).

However, 7 days later, the situation was reversed. The patients receiving placebo had a marked, gradual decrease of muscular tonus following the immediate post-birth period. This was no surprise since we had observed this phenomenon in other untreated Down's syndrome infants prior to starting this study. In part, this decline in tonus parallels the drop in whole blood 5-HI seen in both normal and Down's syndrome patients during the first week of life (Fig. 1-4). It also seems more pronounced in patients with increased bilirubin levels.

The patients placed on 5-HTP, although starting with a poorer tone score prior to beginning the amino acid, had marked increase in tone rating 7 days later. The patients receiving 5-HTP then had tonus scores superior to the patients receiving placebo for the next few months but this was no longer present after 6 months of age. In fact, the effect of 5-HTP on increased tone appears primarily limited to the initial months of life, before 'blood-brain' barrier systems and enzymes mature. After that age, it may have an opposite effect by blocking developed transport systems, receptors, etc.

From 6 months until 2 years of age, the tonus scores of patients on 5-HTP

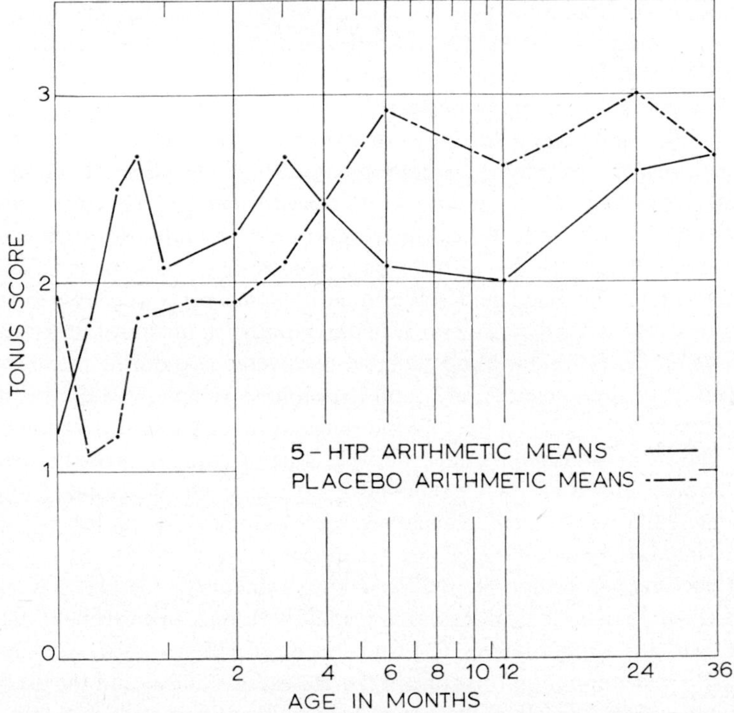

Fig. 3-1. Tonus scores from the neurological examination in patients receiving 5-HTP and patients receiving placebo. The improvement in tonus in the patients receiving 5-HTP noted in the early months of life does not persist and tonus scores in 5-HTP patients are below scores in placebo patients for the remainder of the study.

were depressed below those of the placebo patients. Finally, at the 3-year evaluation, tonus scores in the 2 groups were roughly comparable. Partington *et al.* (1971) tried 5-HTP on Down's syndrome patients at 2 years of age or older and found no consistent effect on tone in these older age groups.

A comparison of the mean mg/kg dosages of L-5-HTP and D,L-5-HTP (Appendix III-4) shows that dosages were highest during the period of poor muscle tonus and were low or moderate during the periods of better tone in 5-HTP patients past 6 months of age.

However, in an individual patient analysis, the role of environmental factors often appears greater than the effect of the amino acid. For example, it was noted that both 5-HTP and placebo patients had loss of tonus following vacations of their parents where the child was left with a mother-substitute, even if this substitute was a presumably devoted grandmother. Although, by 20 months, at the time of the first home psychiatric evaluation, all parents were rating well as accepting and supportive parents (see Chapter 5), this was not completely true in the first few months of the study. The child on 5-HTP with the consistently lowest tonus scores from the first time examined after coming home from the hospital at 4 weeks of age until after 1 year of age, was born into a family where the mother had grown up with a Down's syndrome cousin. The family had a fixed idea of Down's syndrome patients as severely retarded, helpless people who interfere with family life if kept at home. The parent's toleration of this patient was very negative at first; the child deliberately was not handled, etc. When the patient was 13 months old we intervened with a counseling interview with the parents regarding the home environment for this patient. The intelligent, upper middle class parents were able to accept this counseling and did considerably better with the child almost immediately. After we broke the double blind code, we discovered that during the first year, in an effort to overcome the profound hypotonus, we had elevated the mg/kg dosage and the 5-HI levels beyond the range of other patients. In spite of the other factors causing poor tonus, we were able to improve it temporarily by large dosages of 5-HTP. This patient (B-3) is the one who developed a hypsarrhythmic EEG without any clinical seizure phenomena or evidence of other central nervous system damage (see Chapter 7).

All patients had essentially the same environment for the first 3 weeks of life. They were in a Clinical Research Center with the same excellent patient/nurse ratio and same routine. The tonus in most patients improved between the 21st-day examination (last one in the Research Center) and the 30th-day examination (first one when the patient was living at home and brought in as an outpatient). The individual care at home or a natural improvement in muscular tonus at this age may have been a factor, since this improvement included both 5-HTP and placebo patients.

Three patients had marked decline in tonus between the 21st- and the 30th-day examinations. It may have been coincidental, but all 3 of these were the patients from very large families.

One of the interesting things in the individual case analysis was the *placebo* patients (B-4, B-5 and B-13) who had excellent or even hypertonic tonus scores with hyperreflexia during the first 6 months of life. By chance, one of the findings of marked hypertonus and hyperreflexia was recorded completely independently by two separate neurologists. Imagine our wonderment $2\frac{1}{2}$ years later when we found the patient was on placebo and had no systemic disease at the time which might possibly account for the examination. In patients B-5 and B-13, these high scores persisted up until 3 years of age; yet patient B-4, in spite of such an excellent start, became hypotonic with a 'U' response on the Landau at 1 year of age.

Problems related to ataxia were seen in 4 of the patients on the double blind study. Placebo patient B-15 had a sensation that she was falling which developed during a time when the serotonin level dropped to a low of 18 ng/ml (see Fig. 1-1 in Chapter 1). At that time the patient was 32 months of age and spoke a few words. She is said to have complained to the family, "Oh, I'm falling" many times a day. She also began head banging during this period. However, as the 5-HI level returned to its baseline, the patient no longer complained about the sensation of falling and stopped head banging. It is of interest that although this patient walked at 16 months of age, she still had a very broad based gait at the 36-month examination. Patient B-9, who was on the L-form of 5-HTP, developed ataxia when an effort was made to slowly decrease the dose at the end of the study. Two other patients, B-8 and B-10, had 5-HTP lowered rather abruptly and both became ataxic (see Appendix II-1 for details).

These results are based on a system of quantification of tonus scores of resistance to passive movement (Appendix III-3) which may be an illusion. In infants, tone is so affected by handling, time of day, relationship to feeding and other factors that it is almost impossible to be sure of consistent results. However, over 90% of the examinations were performed by the same person (M.C.) and all out-patient appointments were given at the same time of day. Presumably major trends may be correct; minor differences are unlikely to be so. One advantage of working with definitely hypotonic patients, such as Down's syndrome children, is that the magnitude of tonus deviations is often so large as to be unmistakable. In this way (as in many others) this patient group shows in exaggerated, easier to measure modalities, the patterns and trends of lesser affected and normal children.

In summary, there was evidence of increased muscle tonus in patients receiving 5-HTP up to 4 months of age. After that, the effect of the amino acid appeared to have a deleterious effect, if any, until 3 years of age when no effect, either positive or negative, could be discerned. In some cases, the role of other factors (such as environment) appeared to have a stronger effect on tonus than the amino acid.

Postulate Two

Are 5-HTP or placebo environmental effects responsible for earlier onset of the age of walking noted in Down's syndrome patients receiving the amino acid compared to other series in the literature?

Psychomotor milestones of both patient groups are recorded in Appendix III-5. As can be seen, there is no difference in the age of walking between the placebo patients and the patients receiving the D,L-form of 5-HTP. Those receiving the L-form on the average actually walked later than the placebo patients. Probably the most striking thing about these data is the fact that 4 out of 9 (nearly half) of the *placebo* patients walked alone by 16 months of age, a highly unusual finding compared to any previous group of Down's syndrome patients. These 4 patients span a socio-economic group from the poorest patient in the study to the second most wealthy. The only characteristic they seem to have in common is that all 4 patients had mothers very accepting and interested in their child, demonstrating a positive "attitudinal effect"* on their children.

In the other milestones, the placebo patients had equal, and sometimes superior, psychomotor development. Parent report of words spoken (presumably, if exaggerated, it was done equally by both sets of parents) were based on written lists submitted by the parents at the 3 year evaluation. In patient (B-11) with the best speech by parental claim, an understandable 5 word sentence was heard in the office.

In summary, the postulated effect of 5-HTP to lower the age of onset of walking was not verified. Both placebo and D,L-5-HTP patients in Group B walked at 22 months, the same mean seen in the Group A patients. Patients receiving the L-form of 5-HTP walked later, possibly due either to chance or to the toxic effects of the more active form of the amino acid.

The mean age of walking at 22 months, however, remains lower than other series in the literature. 'Attitudinal effects', noted to affect tonus, may account in part for this result.

Postulate Three

Do patients receiving 5-HTP have any difference in frequency or pattern of strabismus compared to patients on placebo?

In the double blind study, 3 patients, B-6, B-13 and B-15 were on placebo and had convergent strabismus. In the case of B-13, it was so severe that an operation was performed. Two patients who received 5-HTP also suffered from strabismus, B-11 and B-17. The parents of B-17 resisted attempts to lower doses of 5-HTP between 2–3 years of age because they reported the strabismus worsened each time the dose was lowered.

With our small series, no correlation between absence of strabismus and 5-HTP administration can be made. Since 5-HTP does not appear to be of value in reducing hypotonia elsewhere in the body after early infancy, it is un-

* Phrase by G. Nellhaus, M.D.

likely that a possible hypotonia component of strabismus would be helped either.

Postulate Four

Are 5-HTP, placebo, environmental effects, selection bias or chance responsible for the low incidence of cardiac symptoms in Down's syndrome patients receiving the amino acid in 5-HTP studies?

There was no evidence in the double blind study that 5-HTP had any effect on cardiac symptoms. Three cardiac patients received placebo and 3 received 5-HTP. There was no effect seen when a patient with severe failure was taken off and then put on the D,L-form of 5-HTP again (Fig. 3-2). Pulmonary artery

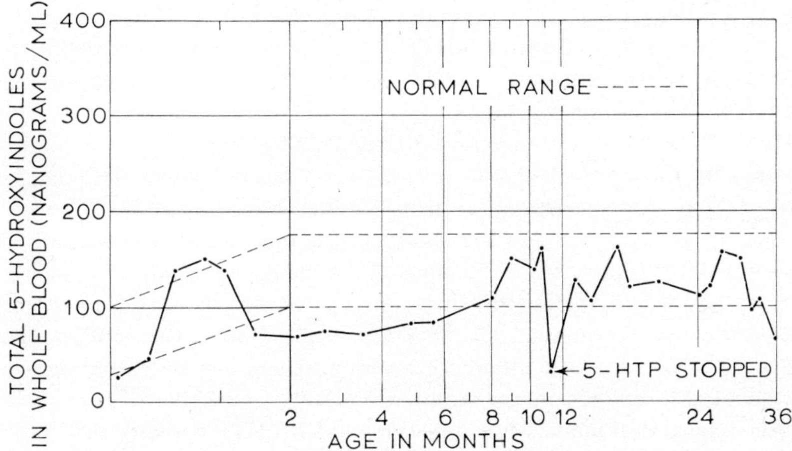

Fig. 3-2. Patient B-8, who received the D,L-form of 5-HTP, had congestive cardiac failure and cyanosis. After the neonatal period, a relatively small dose of 5-HTP failed to keep whole blood 5-HI in the normal range. At 10 months of age (for a 40-day period) the patient was taken off the then unknown solution and the 5-HI level fell markedly. Later 5-HI levels were maintained with 5-HTP.

banding was necessary in this child and in one placebo patient. Apparently the absence of cardiac symptoms in the Group A patients was due to selection and chance factors. Cardiac diagnosis in the Group B double blind patients and in the subsequent C (rare karyotypes of Down's syndrome), D (5-HT+ B6) and E (B6 alone) groups is seen in Appendix II-3.

Postulate Five

Is 5-HTP a factor in the changed pattern of tongue visibility in the patients receiving the amino acid?

Both placebo and 5-HTP parents reported that their children's tongue protrusion lessened with manipulation of the 'medicine'. Patient B-8 was taken off 5-HTP for 40 days when she was 11 months of age. This decision was made by the neurologist and the cardiologist in consultation because the patient's cardiac disease was so severe and it was feared that the research medication might be contributing to the patient's clinical condition. No change in cardiac status occurred during the 40 days when the patient was taken off the then unknown solution. However, when the patient was started back on D,L-5-HTP, the mother reported that the tongue protrusion had improved.

Patient B-10 is a child who had a rare buccal–lingual hypotonia throughout the study. The only time the child suffered from buccal–lingual dyskinesia was during the period when the dose was inadvertently raised too high. In retrospect, this gave a definite impression that the dyskinesia was induced by the elevation of the L-form of 5-HTP. After 3 years of age, an attempt was made to wean the patients off 5-HTP slowly and most parents reported that tongue protrusion became noticeably worse.

However, the same phenomenon was noted in children who were on placebo. In the case of patient B-5, the parents were shocked when the code was broken after 3 years to learn that their child had been on placebo. They cited as their major evidence that the child had the improvement in tongue protrusion with manipulation of the dose of the unknown solution.

In an attempt to answer this question, careful records were kept on both buccal–lingual hypotonia and dyskinesia throughout the study. There appeared to be very little difference between treated and untreated patients. At the 3-year examination, for example, 3 placebo and 3 treated patients had buccal–lingual dyskinesia while 6 placebo and 7 5-HTP patients had buccal–lingual hypotonia.

In summary (in spite of individual cases where the effect of 5-HTP on buccal-lingual hypotonia and dyskinesia seemed so clear) we were unable to substantiate any difference in these modalities between 5-HTP and placebo patients.

Postulate Six

Does administration of 5-HTP affect baseline temperatures in Down's syndrome infants?

A study of the effect of 5-HTP on 8 a.m. temperature in the 5-HTP and placebo patient groups was undertaken during their initial admission to the Clinical Research Center (Fig. 3-3). The patients destined to receive 5-HTP had lower admission temperatures (they also had more depressed muscle tonus). However, (in contrast to increasing tonus) the administration of

5-HTP did not raise their temperatures higher than the placebo group—with the probably chance exception of the 10th-day temperature.

It is concluded that at the dosage level of 5-HTP used in this study, there is no effect on temperature in patients with the trisomy form of Down's syndrome during the first 3 weeks of life.

POSTULATE SEVEN

Is there any objective evidence of increased activity or stress in patients receiving 5-HTP, such as an increase of epinephrine in their urine? How are the catecholamines and their metabolites affected by large doses of 5-HTP?

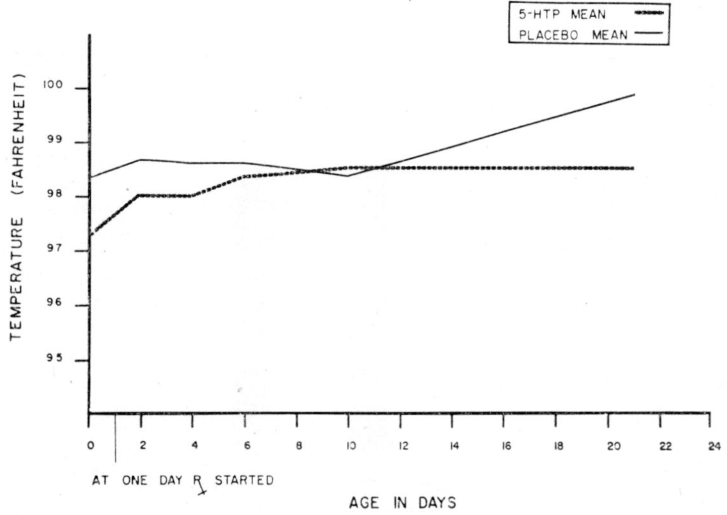

Fig. 3-3. Neonatal temperatures in double blind patients. The patients receiving 5-HTP had lower initial temperatures and maintained a lower level.

The catecholamines (epinephrine [E] and norepinephrine [NE]) and their metabolites (homovanillic acid [HVA], vanilmandelic acid [VMA] and 3-methoxy-4-hydroxyphenylethyleneglycol [MHPG]) were measured in 24-hour urine specimens in the double blind patients at 1, 2 and 3 years of age.

The catecholamines partially share transport mechanisms (Palaic *et al.*, 1967), binding mechanisms (Everett and Borcherding, 1970), enzymes and co-enzymes with the 5-HT pathway. Large doses of 5-HTP will inhibit the synthesis of norepinephrine in rat brain (Johnson *et al.*, 1968). Lowering of one of the enzymes (aromatic L-amino acid decarboxylase) shared by both pathways recently has been demonstrated in patients receiving large doses of L-Dopa, the comparable precursor amino acid of the catecholamine pathway (Dairman *et al.*, 1972). Thus, we would not have been surprised to see lowering of metabolites of the catecholamine pathways in patients receiving large doses of 5-HTP due to transport, binding, enzyme or co-enzyme inhibition.

We had noted increased activity in patients receiving 5-HTP and were aware of the animal literature suggesting that when 5-HTP causes excitation, there is release to receptor sites (and therefore an endogenous depression) of brain levels of norepinephrine (Brodie *et al.*, 1966). However, the CNS presumably affects the adrenal medulla so that the increased activity and stress are associated with increased sympathetic stimulation and increased output of epinephrine in the rest of the animal outside the CNS.

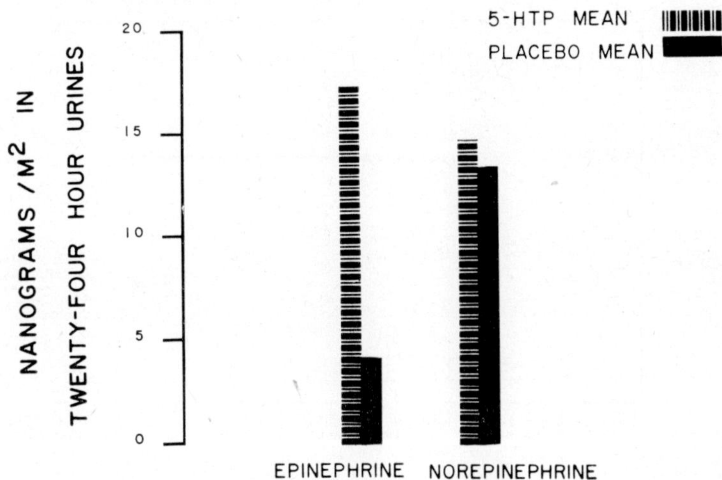

Fig. 3-4. Epinephrine levels in 24-hour specimens were significantly higher in patients receiving 5-HTP compared to placebo patients. No significant difference was seen in norepinephrine levels.

The main difference found between the two double blind groups of patients was in the excretion of epinephrine in the urine (Appendix III-6a). In patients receiving 5-HTP, epinephrine was significantly elevated compared to the placebo patients (Fig. 3-4). It is known that epinephrine levels can be elevated in cardiac patients. A re-calculation was therefore done, omitting these patients. Even with the more limited numbers of patients then available, the difference still remained statistically significant (Appendix III-6b). No difference in effect of catecholamines could be found between the L- and D,L-forms of 5-HTP (Appendix III-7).

Epinephrine is the one catecholamine reported to be depressed in the urine of Down's syndrome patients (Keele *et al.*, 1970; Beauvallet *et al.*, 1970). A depression of norepinephrine has not been found. However, the catecholamine pathway may be generally low in this patient group. Dopamine-beta-hydroxylase (DBH), the enzyme converting dopamine to norepinephrine, has been noted to be depressed in the serum in this patient group (Friedman *et al.*, in preparation).

Objective documentation of increased activity is difficult, even with infants in electronic cribs. The chronic elevation of epinephrine in the urine of the 5-HTP patients could be interpreted as one positive indication of increased activity. Since Down's syndrome patients usually have low levels of epinephrine in the urine, this elevation probably is toward a more normal range. The lack of effect of large oral doses of 5-HTP upon the other catecholamine metabolites is an interesting negative finding.

POSTULATE EIGHT

Is there any difference in the height, weight and head sizes of patients chronically receiving 5-HTP and patients on placebo?

Height and weight data were corrected by parental height and weight figures (Appendix III-8). There was a trend, not of statistical significance, toward lower values in the patients receiving 5-HTP. No differences in head size data could be determined between groups.

Apparently, at doses of 5-HTP used in this study, there is no definite evidence of an adverse effect on growth during the 3 most critical growth years in a patient's life.

POSTULATE NINE

Is the whole blood 5-HI and urinary 5-HIAA significantly elevated in patients receiving 5-HTP compared to placebo patients? What is the relationship of the placebo 5-HI levels to the first visit baseline 5-HI levels of our clinic?

Both blood 5-HI and urinary 5-HIAA levels were significantly elevated in patients receiving 5-HTP compared to patients receiving placebo (Fig. 3-5; Appendix III-6a). As mentioned in Chapter 2, many of the whole blood 5-HI levels were not all 5-HT; occasionally they fell as low as 57% of the total 5-hydroxyindole value in patients on high dosages of 5-HTP; however, most of the samples studied were 85% or more 5-HT. When one patient was taken off 5-HTP for 1 month, her 5-HI value sank precipitously (Fig. 3-2). Platelet counts are reported in Appendix III-9.

The comparison of the mean of the 5-HI values of the 9 placebo patients with the mean 5-HI pattern of all first visits to our clinic (Fig. 3-6) was very interesting and showed the placebo patients had slightly higher 5-HI values than patients not yet affected by the clinic. One placebo patient, B-13, had intermittently elevated levels up to the lower border of the normal range throughout the 3-year period (see Fig. 3-7). During these elevated periods, this patient's 5-HI values were higher than any 5-HI values obtained in a first visit patient of comparable age. Five other placebo patients (B-2, B-4, B-5, B-6, B-12) had values above the trisomy mean for many months and

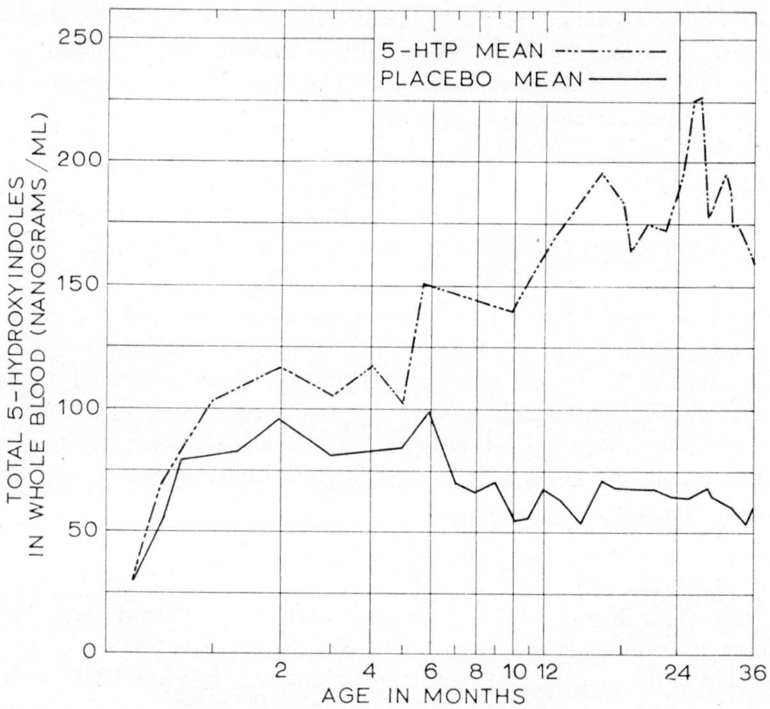

Fig. 3-5. The mean 5-HI values in 5-HTP and placebo patients during this double blind study.

occasional values higher than any first visit patients. Only 1 of the 9 placebo patients (B-16) had a series of 5-HI levels consistently below the trisomy mean.

Additional Clinical Manifestations

Three patients in the study had abnormal mannerisms such as head banging or hitting themselves. In the case of placebo patient B-15, head banging

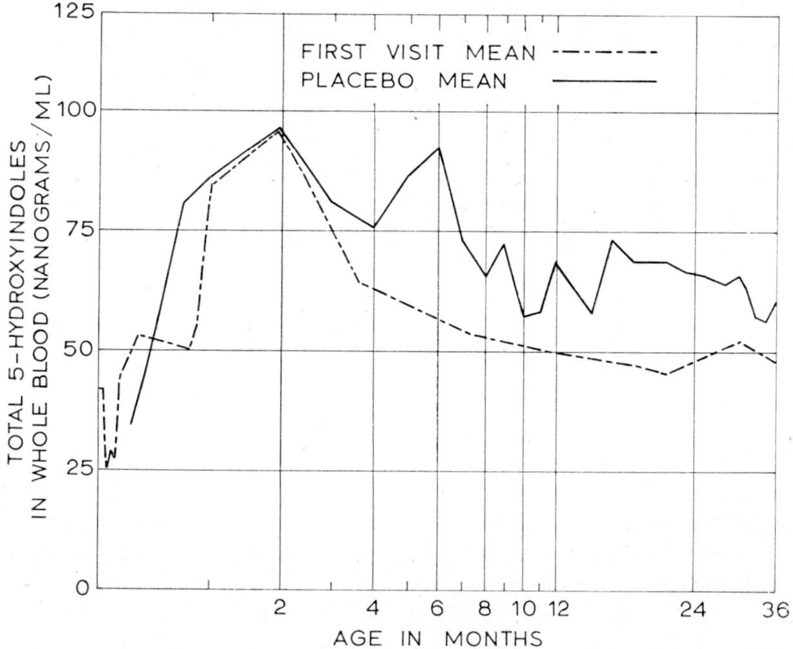

Fig. 3-6. 5-HI mean of all 174 first visit trisomy 21 patients is compared to 5-HI mean of the 9 placebo double blind patients.

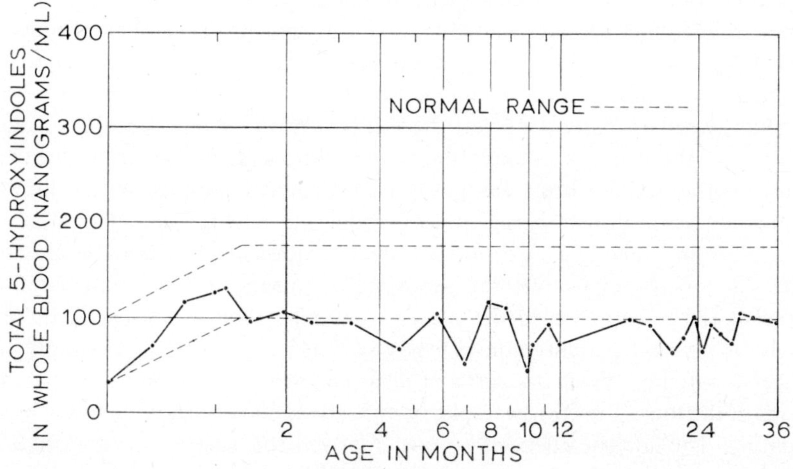

Fig. 3-7. Patient on placebo had a 5-HI value that apparently kept moving up and down, defining the range of the patient's values. Some of the later results near the normal range are higher than any first visit levels ever seen in the clinic.

occurred only at 32 months of age during a 1-month period when the 5-HI level was extraordinarily low (see Fig. 1-1 in Chapter 1). It has not recurred since. Also, in patient B-1, there possibly was evidence that the head banging might be related to the level of 5-HI. The patient was on the D,L-form of 5-HTP for 3 years on this study. At the termination of the study, the child was tapered off the amino acid. However, 6 weeks after 5-HTP was stopped, the child became irritable and began head banging for the first time. 5-HTP was re-started and the head banging stopped. This could have been some type of withdrawal effect which was relieved by re-starting the 5-HTP. A third child in the study to have head banging and hitting herself was patient B-6, on placebo, who had relatively low levels of 5-HI during the second and third year of life when the mannerisms developed.

CONCLUSION

Placebo factors play a major role in many drug studies. Prior to a double blind study reporting no value to thyroid therapy in Down's syndrome patients (Koch *et al.*, 1965), many improvements with thyroid, presumably due to placebo effects, were noted by reliable investigators. We had several hypotheses based on a study of our original patient group and additional ones from the literature. Two of these hypotheses were partially substantiated: an effect on muscle tone in early infancy (early-positive, later-negative) and evidence of increased activity levels (as suggested by epinephrine studies). Many hypotheses were not confirmed by the double blind techniques, in spite of evidence of adequate or more than adequate dosages of 5-HTP even producing toxic CNS effects in some of the patients (see later chapters). Undoubtedly one problem of interpretation is related to the fact that Down's syndrome patients, with so many genes on their extra chromosome, are not an ideal experimental group for testing a single metabolic system. Therefore, the conclusions drawn from both our positive and negative results have a limited interpretation in human beings unless confirmed in other patient groups.

In our clinic, giving a placebo (?magical solution) and helping the parents create a positive environmental atmosphere appears to have contributed to some of the results seen in the placebo patients, such as the elevated 5-HI levels in placebo patients above first visit 5-HI clinic values. Certainly the placebo patients appear to perform at a somewhat higher level than many series in the literature: for example, in milestones such as walking (44% by 16 months). The adverse effect of a poor environment, such as institutionalization, on patients with Down's syndrome has been well established (Centerwall and Centerwall, 1960; Shotwell and Shipe, 1965; Stedman, 1965).

This study reinforces the concept that treatment of the parents is an integral part of the treatment of a retarded child. There is a maxim in medicine "if one cannot help, do no harm". This maxim needs to be applied with great care to parents of retarded children during the critical early months and years of central nervous system development in their child.

SUMMARY

A double blind study comparing 5-HTP and placebo administration in 19 patients, from the neonatal period until 3 years of age, with trisomy 21 revealed the following findings:

1. 32% of the patients had a parent exposed to pelvic radiation prior to conception.
2. 5-HTP affected muscle tonus in the patients, positively in the early months of life and negatively in the years following. An increase in tonus was seen for the first 4 months, then an apparent decrease in tone was seen for the next 2½ years. The 5-HTP dose was elevated during much of this period in a vain effort to improve muscle tone. At 3 years of age, the tonus became equal in both patient groups.
3. An 'attitudinal effect' of the mother on muscle tonus and age of walking was apparent in some cases.
4. 5-HTP, however, had no demonstrable effect on
   a. age of walking
   b. neonatal temperature
   c. buccal–lingual hypotonia or dyskinesia
   d. strabismus
   e. cardiac function
   f. growth patterns.
5. The analysis revealed that abnormal behavior patterns such as head banging and self hitting appeared to be stopped by elevation of 5-HI. This lead needs further investigation.
6. Long-term administration of 5-HTP elevated epinephrine levels in the urine of these patients. No effect on norepinephrine, HVA, VMA, or MHPG could be demonstrated.

REFERENCES

BEAUVALLET, M., BLANCHER, G. and SOLIER, M. (1970) Excrétion urinaire des catécholamines chez les enfants mongoliens à différents âges. *J. Physiol.* **62**, 241.

BRODIE, B. B., COMER, M. S., COSTA, E. and DLABAC, A. (1966) The role of brain serotonin in the mechanism of the central action of reserpine. *J. Pharmac. Exp. Ther.* **152**, 340.

CENTERWALL, S. A. and CENTERWALL, W. R. (1960) A study of children with mongolism reared in the home compared to those reared away from the home. *Pediatrics*, **25**, 678.

DAIRMAN, W., CHRISTENSON, J. and UDENFRIEND, S. (1971) Decrease in liver aromatic L-amino acid decarboxylase produced by chronic administration of L-DOPA. *Proc. Nat. Acad. Sci.* **68,** 211.

EVERETT, G. M. and BORCHERDING, J. W. (1970) L-Dopa: effect on concentrations of dopamine, norepinephrine, serotonin in brains of mice. *Science*, **168**, 849.

FEHLING, C. (1966) Treatment of Parkinson's syndrome with L-Dopa: a double blind study. *Acta Neurol. Scand.* **42,** 367.

FRIEDMAN, L., GOLDSTEIN, M. and COLEMAN, M. (1973) Depression of DBH levels in Down's syndrome. (In preparation.)

HALL, B. (1970) Somatic deviations in newborn and other mongoloid children. *Acta Paediat. Scand.* **59**, 199.

HUNT, N. (1967) The World of Nigel Hunt, Garrett Publications, New York.

JOHNSON, G. A., DIM, E. G. and BOUKMA, S. J. (1968) Mechanism of norepinephrine depletion by 5-hydroxytryptophan. *Proc. Soc. Exp. Biol. Med.* **128**, 509.

KEELE, D. K., RICHARDS, C., BROWN, J. and MARSHALL, J. (1970) Catecholamine metabolism in Down's syndrome. *Am. J. Ment. Def.* **74**, 125.

KOCH, R., SHARE, J. and GRALIKER, B. (1965) The effects of Cytomel on young children with Down's syndrome (mongolism): A double-blind longitudinal study. *J. Pediat.* **66**, 776.

MUENTER, M. D. (1970) Double blind placebo-controlled study of levodopa therapy in Parkinson's disease. *Neurology* **20**, 6.

PALAIC, D., PAGE, I. H. and KHAIRALLAH, P. A. (1967) Uptake and metabolism of (^{14}C) Serotonin in rat brain. *J. Neurochem.* **14**, 63.

PARTINGTON, M. W., MACDONALD, M. R. A. and TU, J. B. (1971) 5-hydroxytryptophan (5-HTP) in Down's syndrome. *Develop. Med. Child Neurol.* **13**, 362.

POSKANZER, D. C. (1969) L-Dopa in Parkinson's syndrome. *New Engl. J. Med.* **280**, 382.

RINNE, U. K and SONNINEN, V. (1968) A double-blind study of L-Dopa treatment in Parkinson's disease. *Europ. Neurol.* **1**, 180.

SHOTWELL, A. M. and SHIPE, D. (1965) Effect of out-of-home care on the intellectual and social development of mongoloid children. *Am. J. Ment. Def.* **68**, 693.

STEDMAN, D. J. (1965) A comparison of the growth and development of institutionalized and home-reared mongoloids during infancy and early childhood. *Am. J. Ment. Def.* **69**, 391.

TRAINER, M. (1971) Ben our special baby. *Baby Talk*, **36**, 6.

CHAPTER 4

Early Behavioral Development in Down's Syndrome

ANN LODGE and PAULA B. KLEINFELD

Background of the Study

There has been increasing appreciation of an apparently critical role of early environmental variables which may facilitate or thwart the full expression of a normal infant's developmental potential. Moreover, the relevance of environmental considerations to the problem of biologically determined mental retardation has been demonstrated by evidence for substantially superior intellectual and social development of home-reared as compared with institution-reared mongoloid children (e.g. Bayley *et al.*, 1971; Butterfield, 1967; Centerwall and Centerwall, 1960; Cornwell and Birch, 1969; Stedman and Eichorn, 1964). The determining impact of early environmental conditions has received further support from the finding of Shotwell and Shipe (1964) that mongoloid infants reared at home during the first 2 years manifested superiority in both intellectual and social development when compared with mongoloid infants placed in an institution from birth, advantages which persisted for at least 3 years after the home-reared children were institutionalized (Shipe and Shotwell, 1965). These trends have contributed to a changing view of the mongoloid child, long seen as a relatively stereotyped, homogeneous entity with limited, though sometimes surprising, capacity for imitation and rote learning, but with a highly predictable future as a moderately to severely retarded individual. Alternately characterized as happy, sociable, affectionate and musical, or as stubborn, aggressive and destructive (Benda, 1969; Belmont, 1971), the child with Down's syndrome is beginning to be appreciated as a more variable being whose developmental course and behavioral characteristics are at least uncertain, if not controversial. Greater longevity probably has contributed to the necessity for revision of these behavioral stereotypes (Belmont, 1971). Belmont concludes from his comprehensive review of historical clinical descriptions as well as more recent behavioral studies of mongolism:

> "There is thus nothing like a widely accepted stereotype of mongoloid behavior, for there is no concensus as to (a) whether an attribute is

unique to mongolism or is shared by other retardates, (b) whether it develops in time or occurs full-blown at the youngest age, or (c) whether it is universal in mongoloids or pertains only to some special intellectually or temperamentally homogeneous subgroup. Yet out of this inconsistent, sometimes irrational, always obscure picture of mongoloid behavior have come the hypotheses of many recent behavioral studies (p. 41)."

There are hopeful indications that these children may have an as yet largely unexplored potential to respond to appropriately timed and sequenced environmental stimulation. Experimental programs utilizing operant conditioning and other approaches to environmental modification and enrichment directed at enhancing intellectual and social competence suggest a heretofore unsuspected plasticity of behavior. For example, as the result of an intensive program of language stimulation and reading training carried out at Sonoma (Ca.) State Hospital, a number of mongoloid children were able to return to the community from the institution and attend special education classes (Rhodes *et al.*, 1961).

In addition to the impact of external environmental conditions, recent biochemical advances have made possible systematic modification of some aspects of the internal environment with the possibility of correcting or normalizing defects at this level. The present study represents a step in the direction of exploring some of the complex interactions resulting from the experimental manipulation of external and internal variables in the trisomy 21 form of Down's syndrome, a relatively well-defined condition where physiological, neurological and chromosomal status can be specified at birth. This chapter will present results of the longitudinal behavioral evaluation of the infants in the double blind study in the areas of mental, motor and social development. These developmental test data constitute a somewhat different emphasis from much of the existing literature on mongolism in its concentration on the early developmental period and emphasis upon generally positive, facilitating external environmental variables. There is relatively scanty longitudinal information available concerning the early development of Down's syndrome infants in normal and comparatively stimulating family settings. Most normative intelligence test data in this area have been obtained from institutional populations where, despite the possibility of biased sample selection favoring the home-reared, there is little doubt that the impact of the underlying organic deficit has been compounded by the effect of maternal and overall stimulus deprivation. Studies based on institutionalized mongoloid populations have generally placed the level of functioning of the majority of Down's syndrome children within the severely retarded I.Q. range (below 35) while studies of home-reared Down's syndrome populations have found that the majority of children tend to function within the moderate range (35–49) and to a lesser extent within the mild range (50–69) of mental retardation (Bayley *et al.*, 1971; Cornwell and Birch, 1969; Dunsdon *et al.*, 1960; Penrose and Smith, 1966; Quaytman, 1953; Tennies, 1943; Wunsch, 1957). It is an axiom of clinical experience that consideration of intelligence quotient alone does not provide sufficient basis for prediction of either educational, vocational or social adjustment. Dunsdon *et al.* (1960) report social competence, as measured by the Vineland Scale (Doll, 1953), to bear a closer relationship to response to education than intelligence quotient alone. Qualities such as interpersonal responsiveness and capacity to relate cooperatively, as well as the less measurable qualities of attention, concentration and purposefulness also appear to play a determining role in overall level of adaptation.

While I.Q.'s above 70 are rarely found in Down's syndrome, a few individual case studies have suggested mild to borderline normal functioning (Zellweger *et al.*, 1968). However, clear chromosomal diagnosis has usually not been established in most instances; results of studies which have included chromosomal analysis have consistently shown that these higher levels of functioning are associated with translocation or mosaicism (Carter, 1967; Shaw, 1962; Finley, 1965).

Many Down's syndrome infants show nearly normal and even promising early development during the first 6 or 7 months of life. However, Down's syndrome infants are characteristically hypotonic from birth, in addition to possessing various physical anomalies, and motor development tends to lag behind the development of sensory and social responsiveness. The relatively normal appearing behavioral development that may be manifest during the first half year characteristically slows down by 9–12 months and perhaps sooner, as evidence is controversial in this area (Carr, 1970; Dameron, 1963; Fishler *et al.*, 1964; Dicks-Mireaux, 1966). Dicks-Mireaux (1966) reports below-average Gesell developmental test performance at 3 months in the areas of motor, adaptive and social behavior. Rate of mental growth in Down's syndrome has been postulated to correspond to a logarithmic progression between chronological age and mental age (MA=log CA) rather than the normal 1:1 ratio (Silverstein, 1966; Zeaman and House, 1962); however, these conclusions are based on data from institutional populations. Meindl *et al.* (1971) found that the mental growth curve for non-institutionalized mongoloid children showed considerable variation from those presented by Silverstein (1966) and Zeaman and House (1962) and rose to a considerably higher level. While selection factors may have affected these findings to some extent, a degree of flexibility in development appears indicated, as well as the possibility of improving intellectual development through the experience of facilitating environmental factors during the early period. Nevertheless, Illingworth (1966) cautions that relatively advanced development during the early months may show a general lack of correspondence with subsequent intelligence levels. This lack of predictability, however, applies to the field of developmental testing in general, where a near-zero or small negative correlation exists between the 1-year test and subsequent intelligence tests. The degree of correlation gradually increases to a point of relative stabilization at about 3 or 4 years, when correlations with later intelligence measures are reported in the literature to range between 0.46 and 0.82, depending on the test interval (Bayley, 1970).

These variable correlations are attributable to many factors including: the plasticity of the infant and his responsiveness to either enhancing or detrimental environmental conditions; the emphasis of the early developmental tests upon the baby's observable sensorimotor and social repertoire which can constitute only a partial (and indirect) mirror of underlying perceptuocognitive processes; the emphasis of subsequent tests of intellectual ability upon verbal skills and abstract concept formation whose relationship to these early behaviors is far from clear; and the changing nature of intellectual processes themselves as well as their questionable definition as a unitary phenomenon. Bayley's observation that intelligence appears to be "a dynamic succession of developing functions, with the more advanced and complex functions in the hierarchy depending on the prior maturing of earlier simpler ones" (1955) suggests the still existing need to identify relevant dimensions as well as the possibly determining role of important underlying reflexes and neuromuscular coordinations.

We do not yet clearly understand the impact upon rate of normal intellectual growth of the timing and other maturational aspects of these simpler abilities. In the case of the mongoloid infant, there may be special developmental discrepancies which complicate the picture even further. The development of the Down's syndrome infant, especially in view of his characteristic motor deficits, may thus be impaired by out-of-phase maturational timing which may interfere with the normal sequencing of the sensorimotor period. Belmont's (1971) evaluation of the existing literature concerned with the mongoloid's pattern of abilities indicates support of areas of relative strength and weakness in sensorimotor skills and perceptual and associative functioning, rather than uniform deficits. While considerable variation is found in the conclusions of such studies, there appears to be evidence for relative weakness in fine motor coordination and control, auditory information processing, production of appropriate vocal responses, and tactile discrimination, while several aspects of visual–perceptual and visuomotor function display relative strength (e.g. Belmont, 1971; Bilovski and Share, 1965; Hermelin and O'Connor, 1961). At present, perceptuocognitive function is one of the most difficult areas to assess in early developmental diagnosis; however, recently developed electrophysiological methods and techniques of computer analysis may provide an important key to identifying relevant underlying neural processes. The comparison of such indicators as the electroencephalogram, heart rate, psychogalvanic skin response, and eye movements may reveal the interaction of different systems in response to stimuli varied in their sensory quality as well as along dimensions of novelty, complexity and meaning. Studies of cortically evoked potentials in response to visual and auditory stimulation have already suggested differences in coding of such information between retarded and normal children (*e.g.* Barnet and Lodge, 1967; Marcus, 1970).

The present study was designed to explore the developmental implications of the finding of diminished whole blood serotonin levels in trisomic Down's syndrome and of reversal of hypotonia in the usually floppy Down's syndrome infant within a short time following 5-HTP administration. The important developmental question arising from this information is whether biochemically sustained improvement in motor tone begun at birth might not enhance both immediate and long-range mental and motor development during a critical early period, as well as have positive long-term effects. An important degree of association between mental and motor development receives theoretical support from many sources, the most notable perhaps being Piaget's stress upon the necessity of adequate progression through the sensorimotor period as a prerequisite for normal cognitive development (1952). Held and Hein (1963) have demonstrated the importance of self-produced movement in visually guided learning, while White and his co-workers, Bruner, and others have stressed the fundamental role of visual skills, sensory feedback and exploratory reaching and manipulation associated with emergent co-ordination of eye, hand and mouth in early development (Bruner, 1967; White *et al.*, 1964). Similarly a learning theory point of view or conditioning approach could predict increased learning based on the larger number of effective transactions with the environment associated with enhancement of motoric abilities through strengthening of stimulus–response bonds.

Considerable variability, however, is noted in the association of mental and motor skills in normal infants and in the rate of their development. While slowness in achieving early psychomotor milestones is frequently associated with subsequent developmental retardation, many bright babies are slow in their motor development and infants showing early motor precocity often fail to display correspondingly high levels of later intelligence.

Stable early personality in temperament and response style also appear to play a role in this interaction, and the quiet, reflective, more sedentary infant is found to be as intelligent as his active, exploring counterpart given a favorable environmental setting (Birns, 1965; Thomas *et al.*, 1963). Subsequent intelligence levels of congenital amputees appear relatively unaffected by the absence of normal sensorimotor experience, with such children showing extraordinary drive and good perceptual–motor development despite some evidence for early language delay (Gouin-DeCarie, 1969; Robinson and Tatnall, 1968). Thus while sensorimotor adequacy and activity may be important for certain aspects of intelligence, there may be significant areas of perceptuocognitive function which are mediated largely through visual and auditory perceptual systems where motor function, other than that directly associated with peripheral sense organs (e.g. eye movements), may play a minimal role.

Bayley reports wide-ranging variation with correlations ranging between 0.24 and 0.78 between Mental and Motor Scale raw scores on her standardization data obtained from infants between 2 and 30 months of age (Bayley, 1965). She proposes that the tendency for such correlations to decrease with age suggests increasing differentiation between mental and motor skills as children attain higher levels of development (Bayley, 1965). However, these higher correlations obtained during the first 10 months of life support the unity of early development and probably important interaction of early sensorimotor development with developing perceptuocognitive function.

The possibility of improving intellectual function through biochemical intervention increases with the rapid development of this field. Again, however, our present lack of precise knowledge as to what aspects of early behavioral development have relevance for subsequent intelligence limits our insight into the specific behavioral goals of biochemical intervention, suggesting the need to proceed cautiously. In diseases such as phenylketonuria and galactosemia where specific metabolic disorders have been identified and corrective dietary treatment instigated in early infancy, the efficacy of such measures in preventing mental retardation have been convincingly demonstrated (Koch *et al.*, 1971a). The finding of depressed levels of whole blood serotonin in children with the trisomy 21 form of Down's syndrome (Rosner *et al.*, 1965) followed by the demonstration of reversal of hypotonia during the neonatal period as the result of administration of the serotonin precursor, 5-hydroxytryptophan (5-HTP) (Bazelon *et al.*, 1967) provided the background for the present study. Pilot work with this treatment suggested improvement in motor tone, increased activity levels and decreased tongue protrusion during the early months of development; findings which raised important questions concerning the possible long-range impact of 5-HTP treatment upon both mental and motor development in the mongoloid child (Bazelon *et al.*, 1967, 1968). To further explore these suggestive results, Partington and his co-workers (1971) examined the effects of a short-term trial of 5-HTP administration upon both serotonin metabolism and various aspects of motor behavior in children with both the trisomy 21 type of Down's syndrome and other forms of mental retardation. No consistent changes could be demonstrated concerning either motor behavior or neurological status in these children, who ranged in age between approximately 2 and 10 years, with the exception of one Down's syndrome subject who displayed marked increase in activity during treatment (Partington *et al.*, 1971). A study by Marsh (1969) of 5 trisomic Down's syndrome children between the ages of 5 and 8 years reported no correlation between serotonin levels and the Vineland Social Quotient (Doll, 1953), a measure closely associated with intelligence. These few studies suggest that the observable effects of serotonin upon behavior appear to lie largely in the motor realm and may be confined to early infancy with no clear indications of possible longer-range effects. However, the increasing evidence for importance of serotonin in normal cerebral processes further emphasizes the importance of ascertaining the role of this neurohormone during the early developmental period.

In view of the ambiguity surrounding the definition of intelligence as well as its relationship to motor skills, and the possible role of serotonin in development, this study is seen as necessarily exploratory and descriptive in nature with little basis for prediction of developmental outcome of 5-HTP administration. Therefore, our approach to the examination of the possible effect of 5-HTP administration upon various aspects of behavioral development is largely an empirical one. A formal prediction of the hoped-for improved development was not made, and two-tailed tests of statistical significance were utilized for analysis of the developmental test data, except where otherwise specified.

Developmental Evaluation

The 20 infants in the study were scheduled for developmental examinations at the ages of 3, 6, 12, 24 and 36 months. No pretreatment observations were made by the psychologist and there was no knowledge as to which treatment group an infant was assigned. The Bayley Scales of Mental and Motor Development (1969) were selected as particularly appropriate in view of their recent standardization on a large representative population sample (Bayley, 1965). With the 2- and 3-year-old infants, the Stanford–Binet Intelligence Scale (Form L-M) was attempted when appropriate. Several previous studies of development in Down's syndrome during infancy and childhood have utilized one or both of these tests which makes possible relevant developmental comparisons for the purposes of this study.

Behavior ratings on a 5 point scale concerned with object orientation and quality of responsiveness, intensity and nature of social responsiveness and activity level were made, utilizing a preliminary research form of the Bayley Infant Behavior Profile developed for the NINDB Collaborative Study (Freedman and Keller, 1963). At the 1-, 2- and 3-year tests, the Vineland Social Maturity Scale was administered during an interview with one or both parents. Additional non-standardized experimental scales concerned with eye–hand coordination, language development and Piaget-type problem solving tasks were also carried out when feasible. An indication of the reliability of the developmental evaluations was obtained through readministration on a blind basis of the Bayley Scales and Stanford–Binet at the age of 3 years by another psychologist (Elizabeth K. Bond) for 17 infants. Approximately half of the first tests at 3 years were carried out by this author (A.L.) while the other half were administered by Miss Bond. The percentage of exact agreement between the two examiners was calculated, item by item, for both the Mental and Motor scales over the range that included both the basal and ceiling performances of the entire subject group at this age. Percentage of agreement was 90% for the mental scale and 92% for the motor scale. For purposes of subsequent data analysis of the 3-year scores, the first test administered was utilized in all but one case where an extremely poor and unrepresentative test was obtained because of the child's behavior.

The parents usually brought the child to the psychologist's office at Children's Hospital for the developmental evaluation, which took from 1 to 3 separate sessions. A few of the tests were carried out in the absence of the parents in the child's hospital room or in the playroom during his stay at the Clinical Research Center. While parents were encouraged to express their own observations and feelings concerning their child's development, they received little specific feedback from the testing during the first year beyond general encouragement, suggestions concerned with recommended stimulation and training, and realistic expectations concerning the development of Down's syndrome children. However, during the second- and third-year evaluations the psychologist usually gave some general interpretation of the child's level and overall pattern of performance in terms of age equivalents. Some counseling was also provided when appropriate concerning parental attitudes and expectations with regard to their child's individual developmental characteristics, as well as information about parents' organizations, educational planning, and community facilities available for retarded children.

Results

To evaluate the results obtained with the Bayley Infant Scales in this study, ratio quotients based on dividing a subject's Mental or Motor Age by his Chronological Age [(MA/CA) × 100] to yield a Mental Development Quotient (MDQ) or a Psychomotor Development Quotient (PDQ) rather than the usual deviation indices (MDI and PDI) were used. These measures were judged more appropriate for use with subjects scoring below 50, and also permit utilization of the Bayley Scales at 36 months, since normative data are available only between the ages of 2 and 30 months (Bayley, 1969; Bayley *et al.*, 1971). It should also be noted that scores obtained with babies who were 2 weeks or more premature were adjusted during the first year of life by substituting the expected date of delivery for the actual birth date.

Mental Development

To assess the effects of 5-HTP administration, results were analysed for the treatment group as a whole as well as for the subgroups receiving the L- and D,L-forms of the drug. In view of the small subject samples involved in this

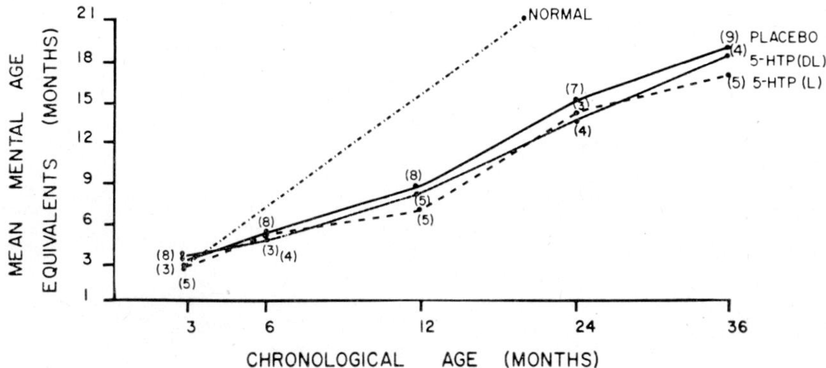

Fig. 4-1. Number of subjects in each age group indicated in parentheses.

study, the results to be reported can only be regarded as suggestive and only tentative interpretations made.

The mean scores of the subgroups and of the total 5-HTP group obtained with the Mental Scale of the Bayley are statistically compared to those of the placebo group by means of the *t*-test (Appendix IV-1). While the placebo group manifests slightly higher Mental Development Quotients (MDQ's) during the first year as compared with either the 5-HTP L- or D,L-subgroups, none of the group comparisons reveal statistically significant differences. Fig. 4-1 graphs the rates of mental growth found with these subject groups from 3 to 36 months, as indicated by their mean Mental Age Equivalents. The relatively increasing deviation from normal test performance with age, generally found during the first 3 years (Dameron, 1963), is illustrated in Fig. 4-2. Here the steepest drop for all groups is seen to occur between 6 and 12 months, then appears to continue to decline less markedly, and is characteristically found to level off at 3 to 4 years of age (Fishler *et al.*, 1964). The significance of the difference in the proportionate drop in MDQ between 3 and 36 months for the placebo group ($\bar{x}$ drop: 54.8 points) as compared with the 5-HTP treated group ($\bar{x}$ drop: 49.3 points) was not significant, according to the results of the Mann–Whitney 'U' test. In general, the results of item analysis (using the Fisher exact probability test) revealed no consistent trends in mental test performance which appeared to differentiate the placebo from the 5-HTP treated group or L- or D,L-subgroups. At 3 months, none of the mental test items were found differentiated significantly between groups; however, the data were suggestive of slightly less social smiling, social vocalization and sustained visual following of objects in the 5-HTP treated group. At 6 months, the data suggested that fewer of the 5-HTP treated infants reached

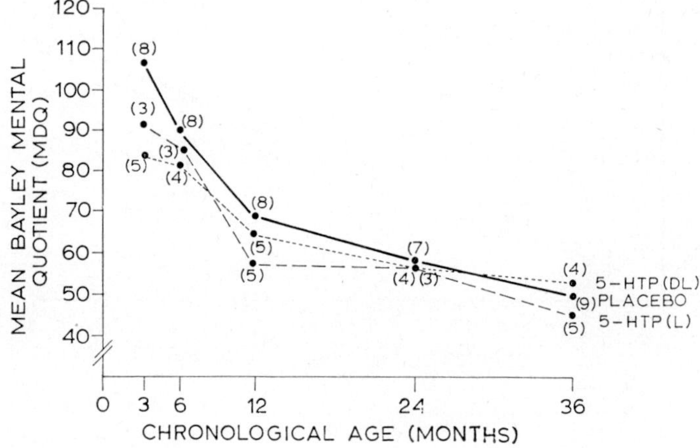

Fig. 4-2. Number of subjects in each age group indicated in parentheses.

for the one-inch red cube as compared with the placebo group; however, the 5-HTP treated group displayed more playful banging of toys. The placebo group showed a trend toward more differentiated vocalization at 6 months (79. Vocalizes 4 different syllables. $p=0.10$). At 1 year of age, the test item 'Looks for contents of box' was failed by a significantly greater number of 5-HTP treated subjects than by placebo subjects ($p=0.02$). On the 2-year test, the placebo group showed some tendency towards more sustained and efficient completion of such tasks as the peg board or filling a cup with cubes, in comparison with the 5-HTP treated group. No mental test items, however, were found to differentiate significantly between the placebo and 5-HTP treated subject groups at either 2 or 3 years of age.

None of the subjects in the 3-year test group passed a sufficient number of Binet items at the 2-year level to yield a scorable test. Children in the overall

5-HTP group passed a mean number of 0.78 Binet items at the Year II level (D,L$\bar{x}$ = 0.50; L$\bar{x}$ = 1.00) while the placebo group passed a mean number of 1.89 items at this level. Analysis of individual items passed revealed no tasks which differentiated between groups on the Stanford–Binet.

Psychomotor Development

Appendix IV-2 presents a comparison of means for the Psychomotor Development Quotients (PDQ) obtained with the various subject groups which are illustrated graphically at the different ages in Fig. 4-3. At 3 months of age, the L- and D,L-subgroups are similar in their motor performance, which falls

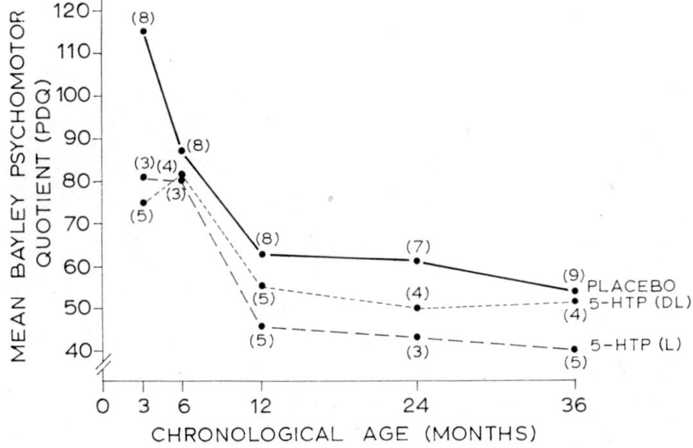

Fig. 4-3. Number of subjects in each age group indicated in parentheses.

significantly below that of the placebo group whose motor test scores are rather high for a mongoloid population ($t = +3.023$; $p < 0.01$). The results of the F-ratio obtained for these two subject groups did not indicate significant differences in their population variances. However, it is of interest that two infants in the placebo group (B-13 and B-14), the only Negro infants in the present study (both premature), received unusually high scores at 3 months on both Mental and Motor scales. Normal Negro infants have been found to show precocious motor development during the first year of life which appears to be associated with increased motor tone (Bayley, 1965). It would be of interest to obtain more information concerning serotonin levels in the Negro population and to examine in detail the early behavioral development of the Negro mongoloid child.

No significant group differences are found at 6 months. However, by 1 year the mean PDQ for the L-group of 45.4 has fallen significantly below the placebo group mean PDQ of 62.8 ($t=2.950$, $p<0.02$), and a similar discrepancy is present at 2 years ($t=2.756$, $p<0.05$). At 3 years, the mean motor performance of the L-group is still below that of either the placebo or the D,L-group (whose mean scores are similar), but the significance of the difference between the mean of the L- and placebo groups now falls between the 0.05 and 0.10 levels of significance ($t=2.165$). The Mann–Whitney 'U' test was used to test the significance of the proportionate drop in PDQ between 3 and 36 months in the placebo group ($\bar{x}$ drop: 60.7 points) as compared with the overall 5-HTP treated group ($\bar{x}$ drop: 32.0 points), as well as with the 5-HTP D,L-subgroup ($\bar{x}$ drop: 23.6 points) and the 5-HTP L-subgroup ($\bar{x}$ drop:

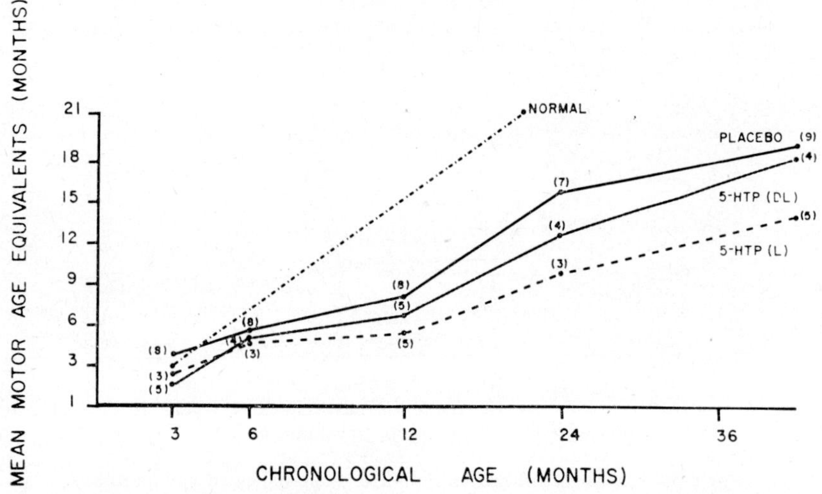

Fig. 4-4. Number of subjects in each age group indicated in parentheses.

40.1 points). Significance of the difference between the overall 5-HTP treated group and the placebo group fell between 0.05 and 0.10 ('U'=12) while the difference between the placebo group and the 5-HTP D,L-subgroup was significant beyond the 0.02 level of confidence ('U'=2) with the placebo group showing a more marked decline from their 3-month level of psychomotor performance in both cases.

Comparative rates of motor growth for the 3 groups are presented in terms of their mean Motor Age Equivalents in Fig. 4-4. Consistently lower performance of the L-subgroup is seen after the age of 6 months, with the tendency of the D,L-subgroup being to perform at an intermediate level and the placebo group demonstrating the greatest competence in motor development. Individual item analysis (using the Fisher exact probability test) of the 3-month Motor test data indicated superiority of the placebo group in comparison with the total 5-HTP treated group in elevation by arms from a prone position ($p=0.05$) and turning from back to side ($p=0.05$). Placebo infants at this age also gave the impression of better-developed head control and the data suggested a trend ($p=0.10$) towards better ability at ulnar–palmar prehension. The 6- and 12-month Motor scales revealed no items which differentiated between the experimental groups. At 2 years there was a trend ($p=0.10$) for placebo group performance to surpass that of infants in the 5-HTP L-subgroup in ability to walk alone, walk sideways, walk backwards, and stand up without assistance from a supine position. No Motor scale items differentiated significantly between placebo and 5-HTP treated infant groups on the 3-year examination.

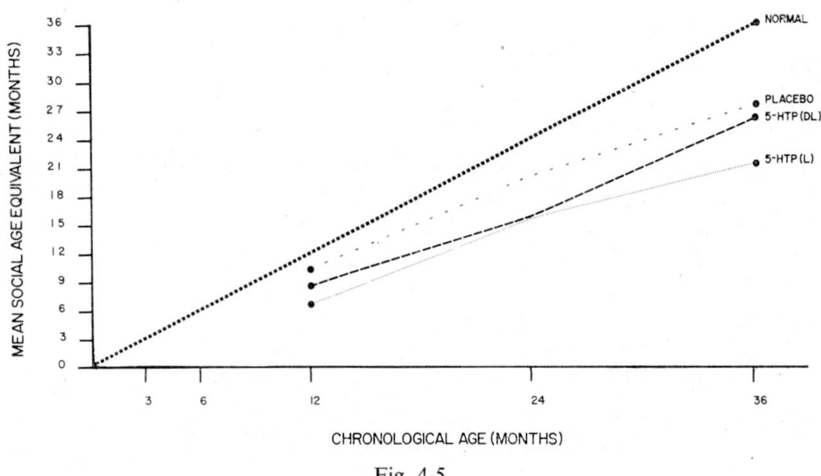

Fig. 4-5.

Descriptive details of Bayley test performance by age for various categories of behavior for the entire Down's syndrome subject group are presented in Appendices IV-3 and IV-4. Despite evidence for some impairment in certain aspects of performance in the 5-HTP treated infants as compared with those receiving the placebo, developmental milestones for most of these home-reared infants irrespective of subject group suggest relatively good development for Down's syndrome infants.

Social Development

Appendix IV-5 summarizes the group means with regard to social development as measured by the Vineland interview. Fig. 4-5 illustrates the rate of social growth for each group as indicated by their mean Social Age Equivalents. Again, a consistent trend toward lowered performance in the L-group as compared with the placebo group may be seen. At 2 years the social adequacy of the overall 5-HTP group is significantly below that of the placebo group ($t=2.510$, $p<0.05$). Appendix IV-6 presents a summary of mean Mental, Motor, and Social Age Equivalents which may be compared with mean Chronological Age for the various groups at each age level. At 1 year, mean Mental, Motor, and Social Age Equivalents appear roughly comparable. By 36 months, these scores show more differentiation, with mean Motor Age substantially lower than either the mean Mental or Social Age Equivalents in the 5-HTP L-group. Mean Social Age is found to be comparatively higher than either mean Mental or Motor Age for all subject groups at 36 months.

At the age of 2 years, the overall 5-HTP treated group received significantly lower Social Quotients than did the Placebo group (Appendix IV-5). However, no individual items on the Vineland were found to discriminate significantly between these 2 groups at this age. Further analysis of the results of the social development data is summarized in Appendix IV-7, which presents mean point scores for the various Vineland behavior categories for the infant subject groups. Inspection of this table reveals that the 5-HTP treated infants were reported significantly less adequate than placebo group infants in the areas of Locomotion ($t=2.612$, $p<0.05$) and Occupation ($t=2.781$, $p<0.05$), with evidence of a trend towards poorer performance in Self-Help General ($t=1.879$, $p<0.10$). Socialization items *per se* did not differentiate significantly between the two groups, despite some suggestion of less adequate social behavior in the 5-HTP treated infants. These findings may suggest that the poorer motor performance (and locomotor competence) of the 5-HTP treated infants, which is also evident at 2 years (Appendix IV-2) may contribute significantly to a lesser overall social competence as well as to other specific categories in the self-help area. The lower score of the 5-HTP treated group in the area of Occupation is perhaps suggestive of a more general deficit in the organization of behavior.

BEHAVIOR RATINGS

The results of behavior ratings carried out during the developmental examinations are presented graphically for the various subject groups in Figs. 4-6 and 4-7. Fig. 4-6 is concerned with the subject's orientation to objects. Speed, intensity and duration of response as well as persistence in pursuit of toys and objects were rated on a 5-point scale where a score of 1 represented a low degree, and 5 a high degree, of the characteristic in question. No significant group differences are found between subject groups in Reactivity and Speed

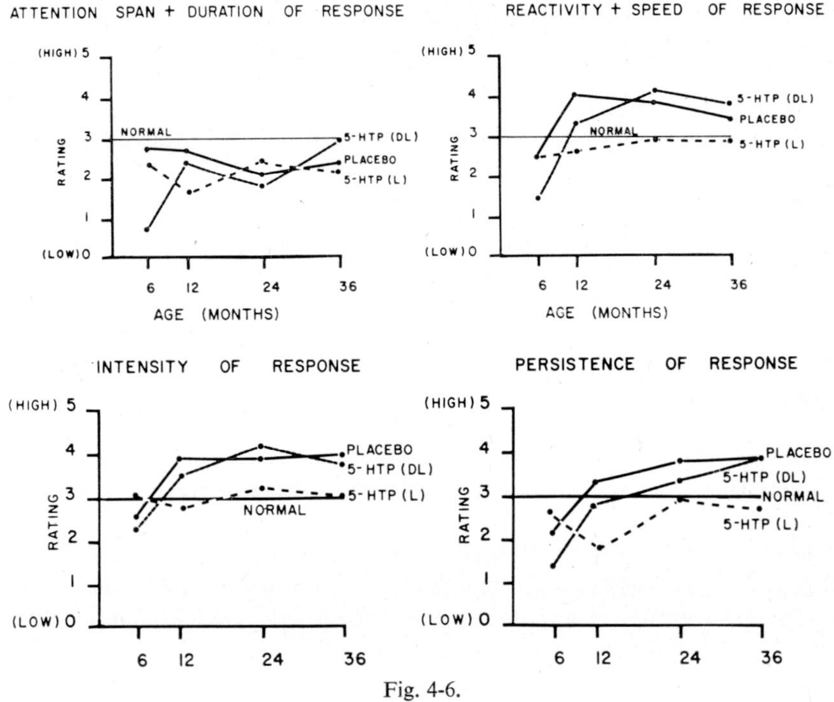

Fig. 4-6.

of Response or Intensity of Response categories. However, a trend is suggested by these data in the direction of longer attention span and duration of response on the part of placebo group infants when contrasted with infants in the 5-HTP L-subgroup at 12 months. Significantly greater persistence of

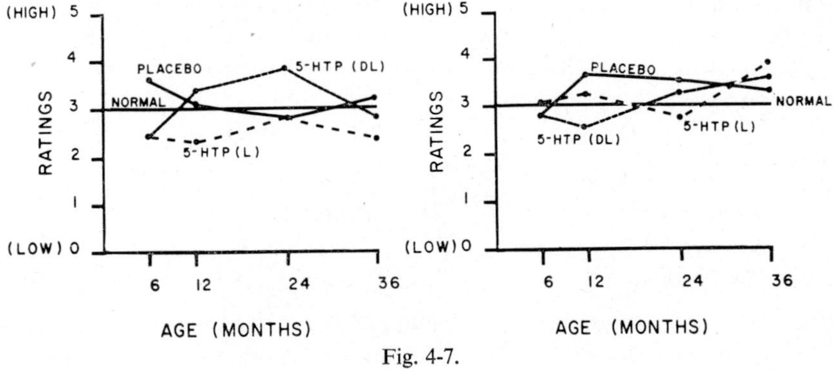

Fig. 4-7.

response at 12 months in the placebo group infants as compared with those in the 5-HTP L-subgroup was also found ('U' = 17, $p = 0.05$).

In the realm of social responsiveness, there is the suggestion of a trend toward lesser intensity of social behavior on the part of the overall 5-HTP treated group at 6 months as compared with the placebo group. The 5-HTP D,L-subgroup then received higher ratings on this variable at 12 and 24 months, while the 5-HTP L-subjects tended to receive relatively low social responsiveness ratings throughout. None of these differences, however, reached statistical significance.

While 5-HTP administration appeared clinically to result in increased activity, ratings of activity level obtained during developmental testing failed to show a statistically significant difference between 5-HTP treated and placebo group infants.

Correlations Between Mental and Motor Performance

The relationship between Mental and Motor Development Quotients as a function of age was examined separately for both the overall 5-HTP treated and placebo groups. These results, using Spearman rank correlation coefficients (r_s), are summarized and compared with Bayley's (1969) normative standardization sample (Appendix IV-8). At the ages of 3 and 6 months, the correlation between Mental and Motor Quotients for both the 5-HTP treated and placebo Down's syndrome groups tend to run higher than those reported for Bayley's normal infant subjects. This discrepancy is even more apparent in the 1-, 2- and 3-year measures where these correlations tend to remain higher for the Down's syndrome groups while dropping off sharply for the normal infants. Both 5-HTP treated and placebo groups manifest relatively lowered MDQ/PDQ correlations at 2 years, while only the 5-HTP treated infants again show a high correlation ($r_s = 0.787$) at 3 years. The implications of these findings are uncertain, due in part to variation in the nature of the test items passed by the normal and retarded infants at the different ages; however, perhaps they reflect relative immaturity associated with less differentiation of function in the Down's syndrome infants.

Relationship of Developmental Data to Physical Variables

Head Size

The relationship of head size to Mental and Motor Development Quotients at 3 years was examined by means of the Spearman rank correlation coefficient for the entire Down's syndrome group. The Psychomotor Development Quotient showed a strong positive correlation with head size ($r_s = 0.482$, $p < 0.05$) while no relationship of head size to Mental Development Quotient was found ($r_s = 0.138$). Since degree of microcephaly has been reported to show a positive correlation with degree of mental retardation, this finding raises some interesting questions concerning the role of psychomotor competence in this relationship.

5-Hydroxyindole Levels

The relationship between 5-HI values, obtained shortly after birth before the onset of treatment and again at the time of testing, and the developmental

TABLE 4-1

Summary of correlation coefficients (Spearman) between 5-HI values and Mental and Motor Scale scores by age for the total Down's syndrome infant group

	3 mths	6 mths	12 mths	24 mths	36 mths
Concurrent measures					
5-HI/PDQ					
n	16	15	18	11	18
r_s	0.010	−0.188	−0.306	−0.143	−0.204
p	ns	ns	<0.20	ns	ns
5-HI/MDQ					
n	16	15	18	11	18
r_s	0.209	−0.211	0.057	−0.125	−0.246
p	ns	ns	ns	ns	ns
Pretreatment 5-HI values					
5-HI/PDQ					
n	13	15	18	14	18
r_s	0.488	0.707	0.078	0.025	−0.115
p	<0.10	<0.01	ns	ns	ns
5-HI/MDQ					
n	13	15	18	14	18
r_s	0.501	0.715	0.041	−0.364	−0.390
p	<0.10	<0.01	ns	ns	=0.10

test results for the total Down's syndrome subject group was studied (Table 4-1). 5-HI values determined at the time of the developmental test appear to manifest a slight increasing negative relationship as a function of age with both MDQ and PDQ; however, no significant degree of association is demonstrated (Appendix IV-9). At 12 months, the correlation of −0.306 between PDQ and the concurrent 5-HI value just misses the 0.10 level of significance by two-tailed test, raising the possibility of interpretation as a trend.

Examination of the correspondence between the pretreatment 5-HI values and subsequent developmental test scores revealed a fairly strong positive relationship at 3 months of age with both MDQ and PDQ ($p<0.10$) and a highly significant positive relationship with both of these scores at the age of 6 months ($p<0.01$). However, at 12 months this correlation is reduced to approximately zero, followed by indications of some degree of negative

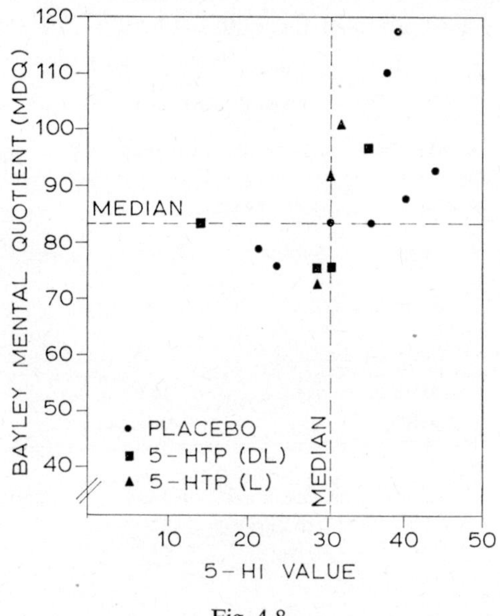

Fig. 4-8.

correlation between pretreatment 5-HI and MDQ which assumes the appearance of a trend by 3 years ($r_s=-0.390$, $p=0.10$). (Figs. 4-8 and 4-9. Also see Appendix IV-9.) The striking early positive relationship between 5-HI levels and developmental test quotients suggests that this biochemical measure at birth may serve as a prognostic indicator of good early behavioral development. However, the subsequent lack of or negative correlation between these measures implies that this relationship may be a relatively short-term phenomenon. The question arises as to what developmental aspects of brain function are reflected in this apparent relationship between the neurological substrate at birth and subsequent measures of behavioral development.

Electroencephalogram

Mean developmental quotients (both MDQ and PDQ) are consistently lower for Down's syndrome infants with abnormal EEG's at 1, 2 and 3 years of age, irrespective of treatment group, although these differences, while indicative of a trend, generally did not reach statistical significance. However, at 3 years, infants showing normal EEG patterns earned an average PDQ of 55.3 while those classified within the borderline to grossly abnormal EEG range had an average PDQ of 45.7 ($t = 1.830$; $p < 0.05$: one-tailed test). Interpretation of this finding is complicated by the fact that most of the

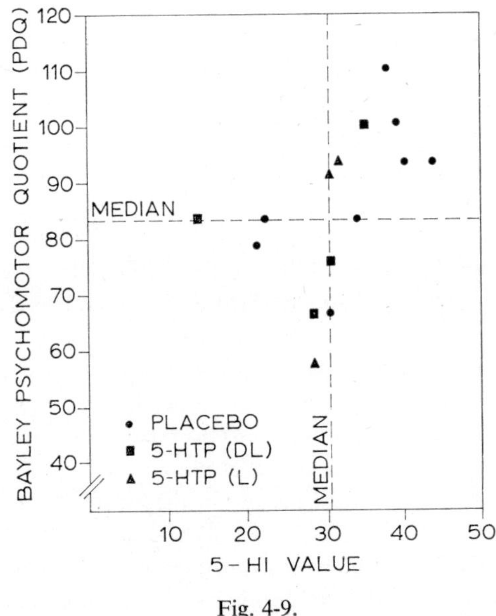

Fig. 4-9.

infants with abnormal EEG's at 3 years were in the 5-HTP treated group and tended to have higher 5-HI values as well (Appendix IV-9). However, it is perhaps significant that the Psychomotor Development Quotient appears either to vary together with, or be adversely affected by, both the effects of 5-HTP treatment and the presence of EEG abnormalities to a greater extent than the Mental Development Quotient.

Discussion

While these diverse findings do not lend themselves to ready interpretation, they do appear to offer some suggestive leads. Evidence for even minimal differences between treatment groups is strengthened by some consistent longitudinal trends, although the small sample size severely limits the generality of the results.

There are several additional factors which complicate the comparison of these results with findings of other studies. These are, briefly: (1) differences in child rearing conditions, (2) heterogeneity of chromosomal diagnoses of other subject groups, and (3) differences in method of computing developmental quotients.

Both the placebo and 5-HTP treated infants in this investigation appear to reflect the considerable impact, in most instances, of being reared within the context of an accepting family who may be highly motivated and unusually hopeful with regard to the child's developmental outcome. In addition to the psychological effect of consistent medication, the parents received frequent and continuing emotional support from medical, psychiatric and psychological staff for 3 years, during the infants' most crucial period of development.

While an untreated control group was not systematically followed as part of this study, Bayley Scale records of a small number of untreated, middle-class, home-reared trisomic Down's syndrome infants at 3, 12 and 36 months were compared with those obtained in the double blind series (Appendix IV-10). Both Mental and Motor mean developmental quotients for the 3-month untreated group ($n=4$) are substantially lower than those obtained with the placebo group at the same age. At 12 and 36 months, however, the untreated and placebo group scores appear much more similar. These comparisons suggest that if there is a substantial 'placebo effect', its impact may be more evident during the early months of development. The potential of intensive early support of families with Down's syndrome infants for enhancing overall development should be further explored. There has been increasing recognition that the manner in which parents are informed regarding the birth of a mongoloid infant and advised concerning early developmental expectations is frequently poor from a psychological point of view. The early rejection that may be fostered by the general lack of medical interest and absence of a comprehensive program of both professional and community support during these earliest years may contribute to significantly lower levels of development in these infants, even when they are reared at home. The proportionately greater decline in placebo group PDQ between 3 and 36 months, as compared with the 5-HTP group, suggests that this measure may be particularly sensitive to environmental variables, especially during the early months of life. In a study of the effects of supplementary 'mothering' given to institutionalized but otherwise normal infants, motor development was found to be one of the variables most profoundly facilitated during the first year (Lodge *et al.*, 1970). In the case of the 5-HTP treated infants, the positive psychological effects of the treatment program may have been counteracted at various stages of development, at least temporarily, by detrimental effects of 5-HTP administration which became evident first in lowered motor tone, followed by lowered Psychomotor Development Quotients, and finally in the development of EEG abnormalities.

While the good general development of the placebo group infants, especially during the first 6 months, does suggest possibly facilitating effects of a supportive treatment program, the possibility of an initial sampling bias in favor of the placebo group cannot be overlooked; however, neither the difference in population variances for mental and motor development quotients nor pretreatment 5-HI levels between the placebo ($\bar{x} = 33.56$) and 5-HTP treated ($\bar{x} = 28.45$) groups were statistically significant. The data from this small sample are insufficient to evaluate possible differences which may be related to sex or race. The high early scores of the two Negro infants in the group have been previously mentioned. There were 4 males in the placebo group and only 2 males in the 5-HTP treated group (both in the L-subgroup), but no consistent performance difference as a function of sex was evident.

The major developmental test finding appears to be that of at least temporary impairment of psychomotor performance, as measured by the Bailey Scales, in 5-HTP treated as compared with placebo group infants. The overall group difference on psychomotor scores is significant at 3 months and 2 years, and suggestive of a trend at 12 months. The groups did not differ in psychomotor test performance at either 6 months or 3 years. While the difference at 3 months of age appears to be primarily attributable to superior placebo group performance, perhaps due in part to selection and environmental variables, the subsequent group differences found appear to reflect relative impairment on the part of 5-HTP treated subjects. Infants receiving the more potent L-form appear to be affected more severely than those receiving the D,L-form of 5-HTP. No significant group differences with respect to mental test performance are found, suggesting some degree of unanticipated dissociation between motor and mental test performance during the early developmental period. These results appear similar to a finding reported earlier with a 5-HTP overtreated infant whose motor development became significantly impaired during the first year, while her mental development remained comparable to that found in other Down's syndrome infants (Bazelon *et al.*, 1968).

The greater degree of relationship of the PDQ, as compared with the MDQ, to several major variables examined in this study, including 5-HTP treatment, EEG abnormality, head size and possibly environmental facilitation is of considerable interest, especially in view of the presumed relevance of these variables to intellectual function. Perhaps psychomotor development is a particularly sensitive index of factors which disrupt the overall organization of behavior. Of course, the differentiation of function represented by items on the Mental and Motor scales is a relative one. The Motor scale places proportionately greater emphasis on bodily control, large muscle coordination, and to a lesser degree finer manipulatory skills, while the Mental scale places more stress upon sensory and perceptual function, vocalization, cognition and the beginnings of abstract problem solving (Bayley, 1969). However, the Mental scale also reflects perceptual motor integration and fine motor coordination to a considerable extent.

There appears to be a puzzling discrepancy between the apparently positive early effects of 5-HTP administration (*e.g.* increased tonus and activity levels) and subsequent slowing of motor development. Because normality of tonus and of motor development have traditionally been considered to vary together, the relationship of upper and lower extremity tonus scores to PDQ was statistically examined using the Spearman rank correlation coefficient (r_s). For the 5-HTP treated infants, a significant positive correlation between both upper and lower extremity tonus and PDQ ($p<0.01$) was found at 1 and 2 years. This correlation was significant ($p<0.05$) at 6 months only for the placebo group, while the 6-month-old 5-HTP treated infants displayed a negative (-0.400) correlation, although this relationship was not statistically significant. The tonus scores of the placebo group infants were generally lower than those of the 5-HTP treated infants during the first 3 months of life, with the exception of somewhat higher pretreatment scores. Thus their relative superiority in psychomotor test performance was manifest despite moderate to severe hypotonia. On the other hand, tonus scores of 5-HTP treated infants appear to have increased during the first 3 months to the mild to moderate range of hypotonia as a direct result of drug administration. This initial apparent improvement, however, was followed by the development of increased hypotonia by 6 months, which was present on the average until 2 years. At the ages of 1 and 2 years, during this period of lower tonus scores, higher tonus scores were significantly associated with higher PDQ's, while this correlation was not present earlier for 5-HTP treated infants. These findings suggest that the relationship between tonus and PDQ, while far from clear, may have differed for placebo and 5-HTP treated infants during the first 2 years. No significant degree of relationship between MDQ and tonus was found except at 1 year of age, when a positive correlation ($p<0.05$) was found for both placebo and 5-HTP treated infant groups.

The relationships are puzzling to interpret and require more detailed examination of cortical and cerebellar involvement of the motor functions affected. These results may represent toxic effects associated with overtreatment since many of the lower scoring infants on the motor development scale also had abnormal EEG's. The possibility of disorganization of overall CNS-regulating mechanisms by high 5-HTP dosages is raised. The question remains as to whether an optimal dosage exists, or whether it may not be possible to thus manipulate one aspect of abnormal biochemical function associated with Down's syndrome and still maintain the functional integrity of the organism. Similarly, the possible developmental effects of other means of raising serotonin levels, such as through vitamin B6 administration, await further investigation. On the other hand, it may be that the characteristically low serotonin levels of a Down's syndrome infant reflect a delicate biochemical balance essential to the overall organization of behavior.

Summary

Comparison of the early development of trisomic Down's syndrome infants who received systematic administration of 5-hydroxytryptophan or a placebo during their first 3 years of life indicates that 5-HTP treatment appears to result in at least temporary slowing of psychomotor development which may begin as early as 3 months but is no longer evident at 3 years. This effect is more marked in infants receiving the L- as compared with the D,L-form of 5-HTP. No effect of 5-HTP administration upon overall mental development was found. However, 5-HTP treated infants showed less adequate social development than placebo group infants at 2 years of age. Possible long-range effects of this treatment await further follow-up investigation.

One interesting result of the present analysis is the finding of a strong positive relationship between pretreatment 5-hydroxyindole levels and both mental and motor development test quotients at 6 months of age. This suggests that the biochemical status of the organism at birth may serve as an important prognostic indicator of early developmental characteristics.

REFERENCES

Barnet, A. B. and Lodge, A. (1967) Click evoked EEG responses in normal and developmentally retarded infants. *Nature (Lond.)*, **214**, 252–255.

Bayley, N. (1955) On the growth of intelligence. *Amer. Psychologist*, **10**, 805–818.

Bayley, N. (1965) Comparisons of mental and motor test scores for ages 1–15 months by sex, birth order, race, geographical location, and education of parents. *Child Devel.* **36**, 379–411.

Bayley, N. (1969) Manual for the Bayley scales of infant development, The Psychological Corporation, New York.

Bayley, N. (1970) Development of mental abilities. In: P. H. Mussen (ed.), Carmichael's manual of child psychology, Vol. 1 (3rd ed.), John Wiley & Sons, Inc., New York.

Bayley, N., Rhodes, L. Gooch, B. and Marcus, M. (1971) Environmental factors in the development of institutionalized children. In: J. Hellmuth (ed.), Exceptional infant: Vol. 2: Studies in abnormalities, Brunner/Mazel, New York.

BAZELON, M., BARNET, A., LODGE, A. and SHELBURNE, S. A. (1968) The effect of high doses of 5-hydroxytryptophan on a patient with trisomy 21: clinical, chemical, and EEG correlations. *Brain Res.* **11**, 397–411.

BAZELON, M., PAINE, R. S., COWIE, V. A., HUNT, P., HOUCK, J. C. and MAHANAND, D. (1970) Reversal of hypotonia in infants with Down's syndrome by administration of 5-hydroxytryptophan. *Lancet*, **1**, 1130–1133.

BELMONT, J. H. (1971) Medical–behavioral research in retardation. In: N. R. Ellis (ed.), International review of research in mental retardation, Vol. 5, Academic Press, New York.

BENDA, C. E. (1969) Down's syndrome: Mongolism and its management (revised ed.), Grune & Stratton, New York.

BILOVSKI, D. and SHARE, J. (1965) The ITPA and Down's syndrome: an exploratory study. *Amer. J. Ment. Defic.* **70**, 78–82.

BIRNS, B. (1965) Individual differences in human neonates' responses to stimulation. *Child Devel.* **30**, 249–256.

BRUNER, J. S. (1967) Eye, hand and mind. Paper presented at the Society for Research in Child Development meetings, March.

BUTTERFIELD, E. C. (1967) The role of environmental factors in the treatment of institutionalized mental retardates. In: A. A. Baumeister (ed.), Mental retardation: Appraisal, education and rehabilitation, Aldine Publishing Co., Chicago.

CARR, J. (1970) Mental and motor development in young mongol children. *J. Ment. Defic. Res.* **14**, 205–220.

CARTER, C. H. (1967) Unpredictability of mental development in Down's syndrome. *Southern Med. J.* **60**, 834–838.

CENTERWALL, S. A. and CENTERWALL, W. R. (1960) A study of children with mongolism reared in the home compared to those reared away from home. *Pediatrics*, **25**, 678–685.

CORNWELL, A. C. and BIRCH, H. G. (1969) Psychological and social development in home-reared children with Down's syndrome (mongolism). *Amer. J. Ment. Defic.* **74**, 341–350.

DAMERON, L. E. (1953) Development of intelligence of infants with mongolism. *Child Devel.* **34**, 733–738.

DICKS-MIREAUX, M. (1966) Development of intelligence of children with Down's syndrome: preliminary report. *J. Ment. Defic. Res.* **10**, 89–93.

DOLL, E. A. (1963) The measurement of social competence, Educational Publishers, Inc., Chicago.

DUNSDON, M. I., CARTER, C. O. and HUNTLEY, R. M. C. (1960) Upper end of range of intelligence in mongolism. *Lancet*, **1**, 565–568.

FINLEY, S. C., FINLEY, W. H., ROSENCRANS, C. J. and PHILLIPS, C. (1965) Exceptional intelligence in a mongoloid child of a family with a 13–15/partial 21 (D/partial G) translocation. *New Engl. J. Med.* **272**, 1089.

FISHLER, K., SHARE, J. and KOCH, R. (1964) Adaptation of Gesell Development Scales for evaluation of development in children with mongolism. *Amer. J. Ment. Defic.* **68**, 642–646.

FREEDMAN, D. J. and KELLER, B. (1963) Inheritance of behaviour in infants. *Science*, **140**, 196–198.

GOUIN-DECARIE, T. (1969) A study of the mental and emotional development of the thalidomide child. In: B. M. Foss (ed.), Determinants of infant behaviour, IV, Methuen, London.

HELD, R. and HEIN, A. (1963) Movement-produced stimulation in the development of visually-guided behaviour. *J. Comp. Physiol. Psychol.* **56**, 872–876.

HERMELIN, B. and O'CONNOR, N. (1961) Shape perception and reproduction in normal children and mongol and non-mongol imbeciles. *J. Ment. Defic. Res.* **5**, 67–71.

ILLINGWORTH, R. S. (1966) The development of the infant and young child: normal and abnormal (3rd ed.), William and Wilkins Co., Baltimore.

KOCH, R., ACOSTA, P. B. and DOBSON, J. C. (1971a) Two metabolic factors in causation. In: R. Koch and J. C. Dobson (eds.), The mentally retarded child and his family: A multidisciplinary handbook, Brunner/Mazel, New York.

KOCH, R., FISHLER, K. and MELNYK, J. (1971b) Chromosomal anomalies in causation: Down's syndrome. In: R. Koch and J. C. Dobson (eds.), The mentally retarded child and his family: A multidisciplinary handbook, Brunner/Mazel, New York.

LODGE, A., HUNTINGTON, D. S., ROBINSON, M. E. and LEWIS, J. (1970) Enhancing the development of institutionalized infants. *Med. Ann. D. C.* **39**, 628–631.

MARCUS, M. M. (1970) Visual evoked responses to pattern in normals and mental retardates. *Clin. Res.* **18**, 206.

MARSH, R. W. (1969) Serotonin levels and intelligence in trisomy-21 type Down's syndrome. *N. Z. Med. J.* **70**, 179.

MEINDL, J. L., BARCLAY, A. G., LAMP, R. E. and YATES, A. C. (1971) Mental growth in noninstitutionalized mongoloid children. *Proceedings 79th Annual Convention American Psychological Association*, 621–622.

PARTINGTON, M. W., MACDONALD, M. R. A. and TU, J. B. (1971) 5-hydroxytryptophan (5-HTP) in Down's syndrome. *Devel. Med. Child Neurol.* **13**, 363–373.

PENROSE, L. S. and SMITH, G. F. (1966) Down's anomaly, Little Brown and Co., Boston.

PIAGET, J. (1952) The origins of intelligence in children (2nd ed.), International Universities Press, New York.

QUAYTMAN, W. (1953) The psychological capacities of mongoloid children in a community clinic. *Quart. Rev. Pediat.* **8**, 255–267.

RHODES, L., GOOCH, B., SIEGELMAN, E. Y., BEHRNS, C. A. and METZGER, R. (1969) A language stimulation and reading program for severely retarded mongoloid children: a descriptive report. In: A. B. Mills (ed.), California Mental Health Research Monograph #11, State of California Dept. of Mental Hygiene.

ROBINSON, M. E. and TATNALL, L. J. (1968) Intellectual functioning of children with congenital amputation. *Clin. Proc. Children's Hosp. D. C.* **24**, 100–107.

ROSNER, F., ONG, B. H., PAINE, R. S. and MAHANAND, D. (1965) Biochemical differentiation of trisomic Down's syndrome (mongolism) from that due to translocation. *New Engl. J. Med.* **273**, 1356–1361.

SHAW, M. W. (1969) Familial mongolism. *Cytogen*, **1**, 141.

SHIPE, D. and SHOTWELL, A. M. (1965) Effects of out-of-home care on mongoloid children: a continuation study. *Amer. J. Ment. Defic.* **69**, 649–652.

SHOTWELL, A. M. and SHIPE, D. (1964) Effect of out-of-home care on the intellectual and social development of mongoloid children. *Amer. J. Ment. Defic.* **68**, 693–699.

SILVERSTEIN, A. B. (1966) Mental growth in mongolism. *Child Devel.* **37**, 725–729.

STEDMAN, D. J. and EICHORN, D. (1964) A comparison of the growth and development of institutionalized and home-reared mongoloids during infancy and early childhood. *Amer. J. Ment. Defic.* **69**, 391–401.

TENNIES, L. G. (1943) Some comments on the mongoloid. *Amer. J. Ment. Defic.* **48**, 46–48.

THOMAS, A., BIRCH, H. G., CHESS, S., HERTZIG, M. E. and KORN, S. (1963) Behavioral individuality in early childhood, New York University Press, New York.

WHITE, B., CASTLE, P. and HELD, R. (1964) Observations on the development of visually-directed reaching. *Child Devel.* **35**, 349–364.

WUNSCH, W. L. (1957) Some characteristics of mongoloids evaluated in a clinic for children with retarded mental development. *Amer. J. Ment. Defic.* **62**, 122–130.

ZEAMAN, D. and HOUSE, B. J. (1962) Mongoloid MA is proportional to log CA. *Child Devel.* **33**, 481–488.

ZELLWEGER, H., GROVES, B. M. and ABBO, G. (1968) Trisomy-21 with borderline mental retardation. *Confinia Neurologica*, **30**, 129–138.

CHAPTER 5

Personality Development in Patients on the Double Blind Study receiving 5-Hydroxytryptophan or Placebo

LEON CYTRYN and LOVISA TATNALL

The original observations concerning the effects of 5-hydroxytryptophan administration on infants with Down's syndrome, described in Chapter 2, suggested several changes. These were: improvement in muscle tone and in appearance due to decrease of tonguing, increase in activity level and vocalization. All these factors seemed significant in their potential to affect positively the parent's acceptance of the child, the child's responsiveness to environmental stimulation and, consequently, an improved parent–child interaction leading to a higher level in personality development. The double blind study seemed well suited to an exploratory study of these assumptions.

Many studies of children with Down's syndrome describe either children in institutions or seen in public hospitals and clinics, often coming from socioculturally deprived segments of our population. Thus, in addition to their innate retardation, they are often subject to considerable environmental deprivation, which makes the interpretation of the findings extremely difficult. In contrast, our group is predominantly middle-class, consisting of intact families with comfortable incomes and above average educational levels (see Appendix III-2). What is most important about this group is the parents' dedication to the child, manifested in their willingness to enroll in a very taxing research program involving a meticulous adherence to a medication schedule and frequent clinic visits, often from a considerable distance. Thus, we felt that we are likely to see children with Down's syndrome reared under rather optimal conditions and any manifest gain in the treated group would reflect a change in the innate factors within the child.

Methods of Procedure

A personality inventory for infants and toddlers was constructed, based on the concepts of coping, developed by Murphy *et al.* (1956), and of constitutional vulnerability developed by Heider (1966, 1971). The inventory consists of forty-five 5-point scales arranged in four groups, namely: Motor Development and Sturdiness, Social Relatedness, Coping and Mastery, and Emotional Integration (Appendix V-1). In addition, a parent rating instrument was devised consisting of four 5-point scales, namely: Acceptance of Child, Maternal Responsiveness, Maternal Initiative, and Maternal Encouragement of Independence (Appendix V-2). Both the children's as well as the parents' rating scales were unidirectional. A rating of 5 was given for optimal behavior and a rating of 1 was given for least desirable behavior. The ratings were given immediately after a two-hour home visit, during which the child and the parents were observed with minimal interference by the observers. Each child was visited twice for each rating, with a one-week interval between visits, by a team of two observers who rated the child independently, following which they adjusted their findings after an exhaustive discussion. In addition, during each visit one of the observers wrote a running description of the child's and mother's behavior, as well as that of other members of the family. These extensive reports permitted us to preserve the richness of relevant behaviors and interaction to an extent impossible with only the use of rating scales. Throughout all rating sessions the raters were unaware of the child's blood 5-hydroxyindole level.

We tested our inventory on six 20-month-old toddlers with Down's syndrome known to have received 5-HTP (not in the double blind study) as well as on three normal 20-month toddlers. These preliminary results indicated a full interrater agreement on 84% of the scales and a 99% agreement within one point difference. The test–retest, after one week interval, indicated a full agreement of 56% and a 97% agreement within one point difference. The inventory also distinguished well between normal infants and those with Down's syndrome and pinpointed specific areas of personality strengths and weaknesses. For instance, it is accepted that the social development in Down's syndrome is usually good; in the area of social relatedness, the differences between the normal toddlers and those with Down's syndrome were much smaller than in other areas.

We then extended the test to the double blind 5-HTP/placebo study. Since our observations were based on repeated home visits, we were forced to eliminate 3 of the 19 children in the project, who lived outside of a 50 mile radius from Washington. Thus, we were left with 16 children. We rated each child at 20 months and again at 32–34 months. Of the 3 out-of-town children whom we could not rate, 1 was given 5-HTP and 2 were given placebo, which left us with 9 treated and 7 untreated patients. Of those, 1 treated patient dropped out of the study and was unavailable for the retesting at 32 months. As mentioned earlier (Chapter 3) one of the treated patients developed infantile spasms with a hypsarrhythmic EEG in the second half of the first year, at which time the administration of 5-HTP was stopped.

Results of Normal/Down's Syndrome Evaluations

In the preliminary phase of our study we tested our methods on several normal children from intact, middle-class families. In comparing the data on those children and the children with Down's syndrome (both in the treated and the placebo groups) several important differences emerged:

1. The normal children showed more discrimination between mother and strangers. They stayed near mother, checking frequently on her whereabouts. They were reserved with the examiners and especially avoided physical contact with them. Some children with Down's syndrome seemed often oblivious of both the mother and the examiners. The majority, however, were quite intimate in dealing with the examiners, showing little reserve or fear.
2. The normal children were more spontaneous and showed more curiosity in new objects. For instance, when we attempted to videotape some of our sessions they tried repeatedly to touch and manipulate the camera. There was much less curiosity in the children with Down's syndrome.
3. The activity of the normal children was more purposeful and goal-directed while the children with Down's syndrome displayed more random activity.
4. The periods of both action and repose were longer and more sustained in the normal children than in the children with Down's syndrome.
5. There was *less* imitation of vocalization in normal toddlers than in their counterparts with Down's syndrome. However, there was more imitation of non-verbal behavior in the normal group. This phenomenon became more noticeable with advancing age.
6. The normal children showed more pleasure in achievement and expressed it vocally as well as in their facial expression. The children with Down's syndrome found more pleasure in adult approval rather than in their own independent activity, indicating a lesser degree of emotional independence and differentiation.
7. Most children with Down's syndrome did not show any evidence of separation anxiety and fear of strangers until 30–32 months. The clinging phase designated by Mahler (1966) as 'rapprochement' did not appear until the age of 32 months and in many did not appear even by the age of 3 years when our study was completed.
8. In the normal toddlers (when seen at the age of 20 months) there was usually evidence of a 'transitional object' such as a blanket or a toy to which the child was very attached (Winnicott, 1953). In children with Down's syndrome there was some emergence of a 'transitional object' in the form of preference for some toys at the age of 32 months. There was, however, seldom a preference for only one specific object and the objects were readily exchangeable.

RESULTS OF 5-HTP/PLACEBO EVALUATIONS IN TRISOMY 21 PATIENTS

We added up all the scores of our 16 patients on the personality inventory, rank-ordered and divided them into a high group (first 8) and low group (last 8). There were more untreated children in the higher group on the total personality score and several of the subscales, but the differences did not reach statistical significance (Figs. 5-1 and 5-2).

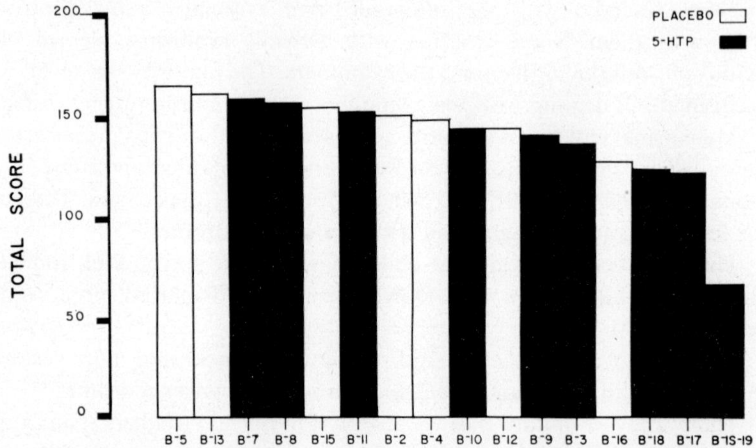

Fig. 5-1. Psychiatric evaluation of the 20-month-old patients in the double blind study. Scores of the placebo and 5-HTP patients on the psychiatric personality inventory.

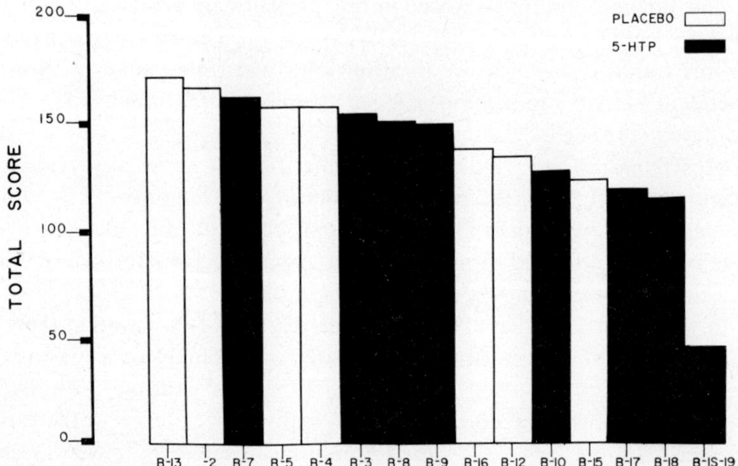

Fig. 5-2. Psychiatric evaluation of the 32- to 34-month-old patients in the double blind study. Scores of the placebo and 5-HTP patients on the psychiatric personality inventory.

We planned to rank-order the mothers as well but this proved impossible since they all clustered between 16–18 points on a 20 point scale.

Discussion

Neurological and psychological studies of the double blind patients have shown that the initially reported changes in children with Down's syndrome receiving 5-HTP were not sustained past the first few months of life. Thus, it should come as no surprise that we found no significant difference in personality development between the two experimental groups. In examining the various subscales of the personality inventory as well as the detailed reports of the home visits, several general factors emerged which distinguished between the children in the high and low groups on the personality inventory scale, regardless of whether they were given medication or placebo.

Firstly, the children in the high group, no matter what solution they were receiving were more *cheerful* and *lively* corresponding to the stereotype of *joi de vivre* which many investigators feel characterizes the majority of children with Down's syndrome (Belmont, 1971). In contrast, the children in the low groups were either more apathetic and lethargic, with a very low energy expenditure, or extremely irritable and restless. This supports the position taken by Engler (1949) and Wallin (1949) about the existence of 3 types of behavior in children with Down's syndrome: (1) the majority who are alert, placid and friendly, (2) a minority who are dull and listless, and (3) some who are negativistic, restless, irritable and very hard to handle.

Secondly, the children in the higher group showed more *awareness* of, and more *responsiveness* to their environment which included people as well as inanimate objects. For instance, although their responses to the examiners varied greatly, their general behavior was definitely affected by the presence of the examiners. In contrast, most children in the low group seemed hardly to notice the presence of two strangers in their house.

The importance of maternal influence on personality development of the child is generally accepted and we looked for the possibility that superior personality development correlated with high ratings of maternal behavior. As stated before, we were unable to rank-order the mothers since they all clustered within the 16–18 point range on a 20 point scale. This makes it likely that all the mothers in our sample performed near the optimal level of maternal behavior. The finding corresponds to our general impression of these mothers and probably relates to the enthusiasm and hope generated by the participation in the research project, as well as the mutual support within this maternal group. Thus, the differences in personality development of the children in our study seemed related not to the differences in maternal handling but rather to differences in innate endowment.

The voluminous literature concerning children with Down's syndrome, by and large, stresses their placidity and good social adjustment. Belmont (1971), in a very illuminating recent review, points to many methodological difficulties often found in behavioral research concerning this group. He found that available evidence does not support the sweeping generalizations often made as to the behavioral uniformity of children with Down's syndrome, although the majority seems to fit the above mentioned stereotype of good social adjustment. Our study, though limited in the number of subjects, seems to support the notion that, in addition to the majority with favorable personality development, there are children with unfavorable personality traits. For reasons mentioned above, we cannot ascribe these differences to maternal influences.

Two other factors logically suggest themselves as a basis for such differences in behavior: (1) the intellectual development, and (2) the presence of EEG abnormalities indicating pathology of the central nervous system over and above those normally found in Down's syndrome. The comparison of our findings with the results of psychological tests (see Chapter 4) indicates a correlation between personality development and intellectual achievement which barely misses statistical significance at 5% confidence at 20 months, but becomes insignificant at 32 months. As to the neurophysiological findings (see Chapter 6), there are more EEG abnormalities in children with less favorable personality development (lower group) than in those with higher personality scores, but the difference failed to reach statistical significance.

The involvement in the double blind study was most fruitful in generating our interest in at least 3 areas of personality development of children with Down's syndrome: (1) assessment methods of personality development of infants and toddlers, (2) attachment patterns in children with Down's syndrome, and (3) emotional disturbances in young children with Down's syndrome. Studies are now in progress attempting to follow the leads evolved in all these areas.

Summary

Personality development of the children with Down's syndrome in both groups (5-HTP and placebo), as well as the behavior of their mothers, were evaluated following two extensive home visits at 20 months of age and two home visits at 32 months. The instruments used were: (1) a specially constructed personality inventory, and (2) a specially constructed maternal behavior scale. The placebo children as a group tended to have higher personality scores but the difference failed to reach statistical significance. There was also no significant difference in maternal behavior in both groups. The features which differentiated between the high and low scoring children (regardless of group assignment) were: cheerfulness, spontaneity, awareness of, and responsiveness to the environment.

REFERENCES

Belmont, J. H. (1971) Behavioral studies in mongolism. International review in research in mental retardation, Vol. 5, N. R. Ellis (ed.), Academic Press, New York.

Engler, M. (1949) Mongolism, Williams and Wilkins, Baltimore.

Heider, G. M. (1966) Vulnerability in infants and young children: a pilot study. *Genet. Psychol. Monogr.* **73**, 1.

Heider, G. M. (1971) Factors in vulnerability from infancy to later age levels. Exceptional infant, J. Hellmuth (ed.), Brunner-Mazel, New York.

Mahler, M. D. (1966) Notes on the development of basic mood, the depressive effect. Psychoanalysis—A general psychology, R. M. Lowenstein, L. M. Newman, M. Schurr and A. J. Solnit (eds.), International Universities Press, New York.

Murphy, L. B. (1956) Personality in young children, Vols. I and II, Basic Books, New York.

Wallin, J. (1949) Children with mental and physical handicaps, Prentice Hall, New York.

Winnicott, D. W. (1953) Transitional object and transitional phenomena. *Int. J. Psychoanal.* **34**, 89.

CHAPTER 6

EEG and Sensory Evoked Potentials

ANN B. BARNET and BETTY L. SHANKS

This chapter examines some of the effects of orally administered 5-HTP on the central nervous system of infants with Down's syndrome as reflected in the electroencephalogram. Brain responsiveness to sensory stimulation was measured using computer-averaged auditory and visual evoked potentials. The EEG was examined for its general characteristics, *e.g.* voltage, frequencies and organization, and for evidence of paroxysmal activity. Characteristics of the EEG sleep cycle were noted. The sample considered in this chapter consists of the Group B (double blind) patients. Recordings were made before the patient started on the treatment regimen, *i.e.* between 2 and 9 days of age, and at 6 months, 1 year (12–14 months), 2 years (24–26 months), and 3 years (35–37 months) of age. A blood sample was drawn for determination of 5-HI, usually within a day or two of the recording.

Recordings were performed at the child's customary nap time which was chosen after conversation with the mother or nurse. The mother was usually present during the testing of the older children. No sedative was used. Most of the data were obtained during sleep. The usual recording period was about 1½–2 hours. Electrodes were applied to the scalp with bentonite and collodion in positions Cz, Oz, T_5 and Fz (10–20 system, Jasper, 1958) with combined mastoids used for the reference electrode and a forehead electrode for ground. Electrode resistances checked at the beginning and end of the session were $< 10{,}000$ ohms. Supra- and infraorbital electrodes were used to monitor eye movements (electrooculogram, EOG) and submental electrodes to record chin electromyogram (EMG). Recording was done in a sound attenuated test booth. A Grass polygraph was used for amplification and to obtain a polygraph recording. A CAT (Computer of Average Transients) was used for on-line averaging of evoked responses. The averaged response was displayed using an X–Y plotter. The EEG was also tape recorded for later data processing. A Packard Bell 250 General Purpose Digital Computer was used for off-line analysis of re-runs of the tape-recorded session. A one-second analysis time was usually used (stimulus + one second) but other intervals were sometimes examined.

In the standard test situation, after calibration and stimulus checks were performed, the EEG without stimulation was recorded. This was followed by auditory stimulation consisting of clicks at an intensity of about 65 dB above voluntary threshold for normal hearing adults. The clicks were produced by the amplified 0.1 msec output of a 161 Tektronix pulse generator. Sets of 100 clicks were presented from a loudspeaker mounted 30 inches from the child's head. Interstimulus interval (ISI) was either 2.5 seconds or irregular with an average ISI of 4 seconds. After several sets of clicks, visual stimuli were presented using a Grass (PS-2) photo-stimulator at the G-16 setting, masked down to a 2-inch diameter circular aperture which was positioned at the infant's midline and 11 inches from his eyes. White and red-orange light flashes were presented. Low steady background illumination in the test booth was used so that an observer could record the child's behavior. Judgements as to the baby's state were made using both behavioral and polygraphic criteria.

Background (Unaveraged) EEG

Pretreatment EEGs in the Neonatal Period

The EEGs in the neonatal period showed the general characteristics reported by Parmelee *et al.* (1967), Ellingson (1958), Anders *et al.* (1971) and others for normal neonates. All subjects showed cyclic alterations in EEG activity and eye movements during sleep (Aserinsky and Kleitman, 1955; Roffwarg *et al.*, 1966). Neonatal 'quiet' sleep showed two basic patterns, the first characterized by continuous slow 1–3 Hz waves, sometimes with very low amplitude 14–16 Hz activity superimposed, and the second by slow wave bursts, 3–8 seconds in length interspersed with 3–5 seconds of EEG flattening (*tracé alternant*, Dreyfus–Brisac, 1964). During both slow wave and *tracé alternant* sleep the electrodes near the eye recorded little or no activity and the chin electromyogram recorded tonic muscle activity. EEG sleep activity shifted periodically to a third pattern characterized by continuous, mixed, relatively high frequencies and low voltage. During this type of EEG, sharp deflections corresponding to rapid eye movements (REMs) appeared on the EOG recording and tonic submental EMG activity decreased. The characteristic EMG and EOG helped differentiate the stage REM (d-sleep, paradoxical) sleep from waking and very light sleep which show a similar EEG, but increased EMG and variable eye movements. Long transitional periods were frequent during which all the criteria for a single sleep stage were not met. These have been termed 'indeterminate sleep' (Anders *et al.*, 1971). Appendix VI-1 gives the percentage of time spent in each sleep stage.

Mongoloid neonates as a whole showed infrequent eye movements during stage REM sleep. Several factors besides the paucity of eye movements during REM sleep make precise grading of neonatal sleep stages in Down's syndrome patients difficult, *e.g.* the frequently very low tonic EMG activity even in non-REM sleep and the lengthy transitions between sleep stages. Single evoked responses to auditory and visual stimuli were often prominent against the relatively low voltage (10–70 μV) background. None of the neonatal records showed spikes or other abnormal paroxysmal activity. The 5-HTP treated and the placebo groups did not show differing characteristics in their pretreatment EEG recording.

Background EEG Characteristics during Treatment

The most striking finding was an increase in the incidence of EEG abnormality in the treated groups over the placebo group. Findings for each group are graphed in Fig. 6-1. Ratings for individual subjects are shown in

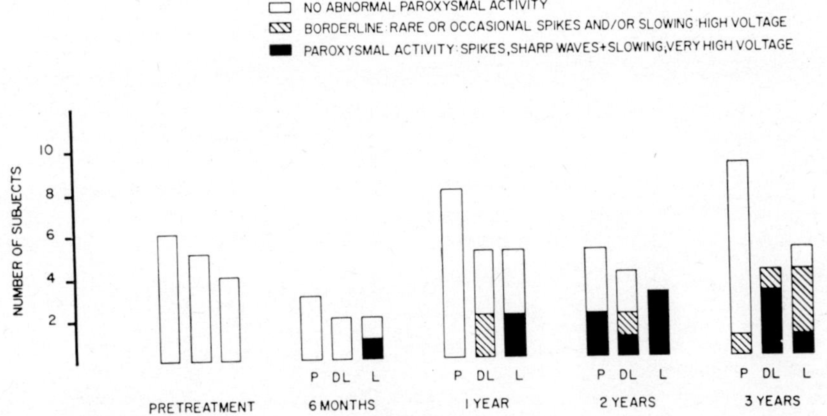

Fig. 6-1. EEG abnormalities in patients treated with placebo (P), D,L- and L-forms of 5-HTP.

Appendix VI-2. Statistical analysis based on those subjects who had recordings at 1, 2 and 3 years of age showed a significantly greater number of borderline abnormal and abnormal records in the treated subjects, *i.e.* of 21 records from 7 treated subjects, 16 showed signs of abnormality, while only 2 of 15 records obtained from 5 placebo infants were abnormal ($\chi^2 = 11.429$, $p < 0.01$). In the treated groups the number of abnormal records increased with age, *e.g.* at 6 months, 1 out of 6 recordings was abnormal whereas 8 out of 9 recordings at 3 years were abnormal. In the placebo group the first abnormal record ap-

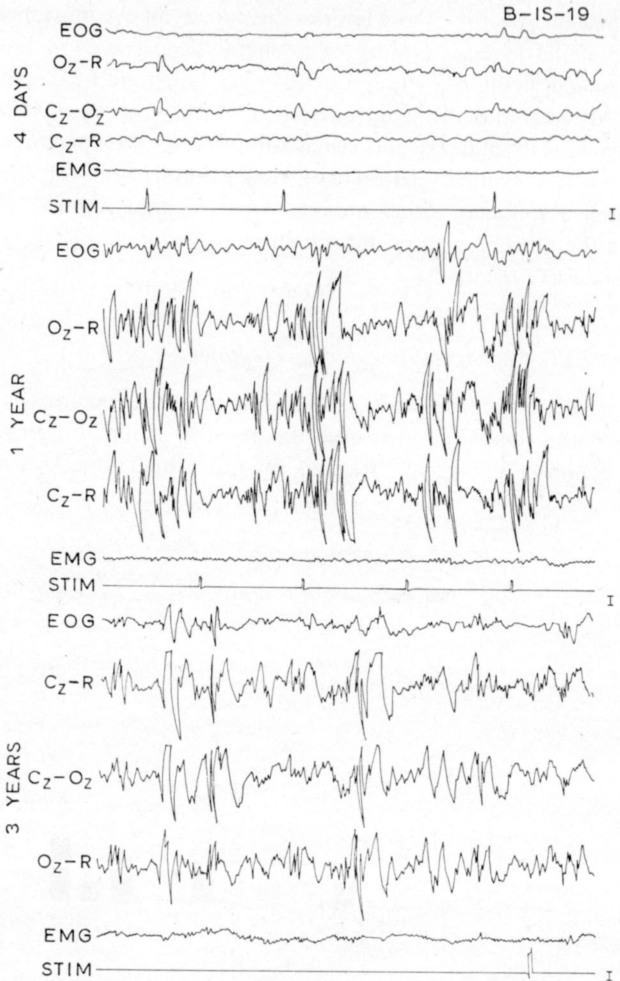

Fig. 6-2. Hypsarrhythmic EEG in a patient (B-IS-19) on L-5-HTP. Calibration: 50 μV.

peared at 2 years. In contrast, in the 5-HTP patients abnormalities appeared in the 6- and 12-month subjects as well.

EEGs characterized by hypsarrhythmia were seen in two 5-HTP treated patients. The hypsarrhythmic pattern consists of mountainous irregular slow waves interspersed with spiking, with periods of near-electrical silence sometimes occurring between high voltage bursts (Gibbs and Gibbs, 1952). Seizures of the infantile spasm type began at 4 months in 1 patient receiving L-5-HTP (B-IS-19) and persisted despite discontinuation of L-5-HTP.

Numerous EEGs showed hypsarrhythmia (Fig. 6-2). Page 124 describes the clinical course of this child. Patient B-IS-19 was the only patient in the double blind group who developed seizures. A hypsarrhythmic EEG without clinical seizures was found in one patient on D,L-5-HTP (B-3) when she was 2 years old but not before. By 3 years the recording of B-3 showed extremely high voltage with paroxysmal slowing and rare spikes. Page 48 describes the clinical course of this patient. No hypsarrhythmic EEG without a history of clinically evident seizures has been previously reported to our knowledge. Hypsarrhythmia as a presumptive toxic effect of 5-HTP is further discussed in Chapter 7.

The EEGs, other than those showing hypsarrhythmia, which were classified as abnormal showed two or more of the following characteristics: extremely high voltage with paroxysmal slowing, occasional to frequent spiking, spikes or polyspikes and slow waves, or runs of high voltage sharp waves (4–8 Hz). Records classified as borderline showed one or more of the following: occasional spikes, high voltage sharp waves or paroxysmal slowing. Fewer EEG abnormalities in both treated and placebo groups were seen in the 6-month and 1-year records than in the 2- and 3-year records. The 3 recordings from different placebo subjects which were abnormal or borderline showed paroxysmal slowing and occasional spikes. The incidence of EEG abnormality in the placebo patients is similar to that reported in the literature for mongoloids (Ellingson, 1973; Ellingson *et al.*, 1970). Because midline electrode placements were used, we do not have information on hemispheric asymmetry or focal abnormality for most of the patients.

The 5-HTP treated children showed other EEG characteristics which differentiated them from both the placebo subjects and normal children. These variants, although clearly drug-related, were not classified as abnormal. Fifty-eight per cent of the 5-HTP recordings (6 subjects) showed fast activity in the 20–26 c/sec range while only 13% (2 subjects) of the placebo recordings showed similar activity. The effect was age as well as drug dependent (Fig. 6-3). Gibbs and Gibbs (1952) reported that 20% of normal 1- and 2-year-olds and 10% of normal 3- and 4-year-olds show 20–30 c/sec spindles which appear to be similar to those of the mongoloid group. Another EEG characteristic of many of the 'treated' children was an excess of extremely high voltage irregular mixed and fast and slow activity during sleep. This effect showed a tendency to be more pronounced in those children with high whole blood HI levels (Fig. 6-4). The long periods of EEG activity of this sort which were seen in the 5-HTP children during afternoon sleep are not characteristic of afternoon sleep of either normal children or untreated mongoloids. For purposes of sleep stage classification, the high voltage fast and slow activity was called stage 3 or stage 4 sleep. Other differences in organization of sleep patterns are discussed in the next section.

EEG Sleep Patterns during Treatment Period

EEG sleep patterns were classified by the system of Rechtschaffen and Kales (1968) for the records of patients 6 months and older. The percentage of time in each sleep stage is summarized in Appendix VI-1. During the EEG recording sensory stimuli were presented which undoubtedly affected the sleep cycle; comparisons with findings obtained during undisturbed sleep must be made with this in mind (Murray and Campbell, 1971).

The mongoloids in the 5-HTP and placebo groups, however, showed differences in sleep patterns which appeared to be related to their treatment status. The 5-HTP treated group and especially those subjects with high whole blood 5-HI levels had periods of very high voltage (500–700 μV or greater) mixed slow (<3/sec) and faster (4–5/sec) activity without sleep

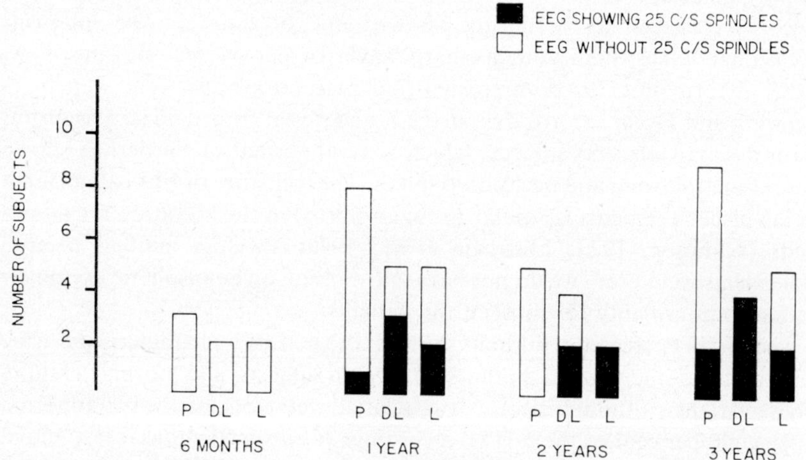

Fig. 6-3. Fast EEG activity in placebo (P) and 5-HTP patients with Down's syndrome.

spindles (see Fig. 6-4). In some subjects this type of EEG lasted for nearly the whole recording except for brief times at the beginning and end of the recording of stage 1 or stage 1 with 'hypnagogic' waves (Kellaway, 1952) which are associated with drowsiness in young children. An increase in slow wave sleep with the administration of 5-HT or its precursors has been reported in laboratory animals (see review by Jouvet, 1969). However, other investigators found no increase in slow wave sleep when 5-HTP was administered to normal adults (Mandell and Mandell, 1965; Wyatt *et al.*, 1971). 'Sleep spindles' were observed more often in the placebo than in the treated subjects, but all patient groups showed less spindling activity than seen in normal children of the same age. Diminished amount and reduced amplitude of sleep spindles

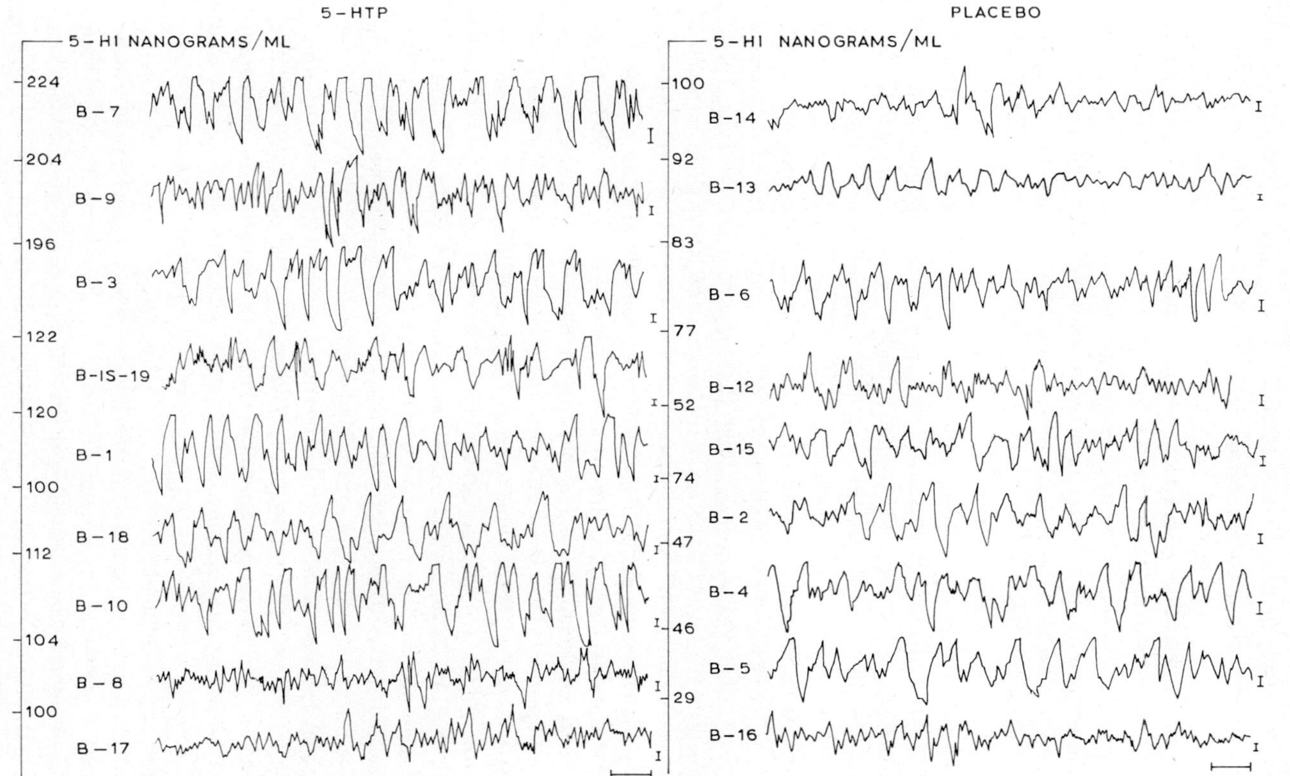

Fig. 6-4. Segment taken from that portion of the EEG, which showed the highest amplitude during the recording at 3 years of age of placebo and 5-HTP treated subjects. The value for whole blood 5-HI (ng/ml) which was drawn within a day of the EEG, appears to the left of each tracing. Lead is Cz (vertex) referred to joined mastoids. Calibrations: 50 μV, 1 sec.

have been reported in mentally retarded subjects including mongoloids (Rhode *et al.*, 1969; Feinberg *et al.*, 1969). Excessive spindles with very high voltage have been seen in some mentally retarded children (Gibbs and Gibbs, 1962).

There were differences by treatment group in the amount of stage REM sleep. At 6 months and 1 year the 5-HTP groups showed less REM sleep than the placebo group. The average percentage REM of the placebo group was within normal limits at 6 and 12 months. However, at 2 and 3 years the amount of stage REM in the placebo subjects was also reduced. The relatively high percentage of stage REM in the D,L-5-HTP group at 2 years was accounted for by 1 subject (B-1) who showed 78% REM while the other treated patients at this age showed none at all. Eye movements tended to be sparse and the characteristic clusters of movement briefer than normal. This was true of both the treated and untreated children; however, the production of eye movements in both groups may have been affected by the sensory stimulation the children were receiving. Decreased amounts of stage REM and decreased rapid eye movements during REM sleep have been previously reported in mongoloids as well as in other mentally retarded subjects (Goldie *et al.*, 1968; Feinberg *et al.*, 1969; Petre-Quadens and DeGreef, 1971).

Administration of 5-HTP *increases* the amount of REM sleep in normal adults (Wyatt *et al.*, 1971; Mandell and Mandell, 1965). There are 3 reports describing the effects on sleep of the administration of 5-HTP to mongoloid infants for varying periods (Bazelon *et al.*, 1968; Lee *et al.*, 1969; Petre-Quadens and DeGreef, 1971). The patients of Bazelon *et al.* and Lee *et al.* showed no clear 5-HTP-related effect on their sleep patterns. However, the preliminary report of Petre-Quadens and DeGreef indicated an increase in REM with 5-HTP in 1 mongoloid child who was started on 5-HTP at 10 months of age. In 2 younger mongoloid infants (4 and 8 weeks old) and in a 4-year-old mongoloid, the effects of 5-HTP on REM were less evident. Besides dose and species differences, there may be age-related differences in the effect of the drug. The duration of drug administration may also be a factor, *i.e.* there may be an acute effect on sleep patterns which diminishes with long-term administration. The numbers of children reported are small and the normal variability in sleep patterns is large. Information on important circadian factors such as feeding and napping schedules, time of day and length of recording has usually not been specified in reports; this makes comparisons of results difficult.

Several sleep patterns occurred in the Down's patient which could not be categorized in the usual way. These 'indeterminate' states are seen more often in abnormal subjects than normal and may be an index of abnormalities in cerebral maturation (Prechtl *et al.*, 1969; Dreyfus-Brisac and Monod, 1970). In many subjects there were 3- to 5-minute periods, often in conjunction with an

REM period, during which the recording showed EEG and EMG characteristics of stage REM but no rapid eye movements. These periods seemed to be an incomplete form of REM sleep. In 1 patient on D,L-5-HTP (B-1), 'incomplete REM' occupied 33 to 44% of total sleep time (except at 2 years). B-1's record at 2 years of age had 78% REM stage and no 'incomplete REM'.

Often at sleep onset but also at other times there were periods when the recording showed a low voltage irregular (stage 1) EEG, with tonic muscle activity and variable phasic muscle activity and eye movements. Sometimes this type of recording merged with the 'incomplete REM' described above. A few low voltage spindles or a body movement might herald stage 2 sleep or arousal; but often the polygraph returned to the indeterminate, mixed, or transitional phase described above. Studies are now in progress in our laboratory in which all-night EEG recordings during undisturbed sleep are being made on 3-year-old mongoloid children treated with 5-HTP since birth and on untreated children given 5-HTP for the first time at 3 years of age. These studies will throw more light on the effects of both chronic and short term 5-HTP administration on the sleep patterns in mongoloid children.

Visual and Auditory Evoked Potentials

Average sensory evoked EEG responses were obtained to every set of stimuli presented in each patient. Single evoked responses could sometimes be detected in the unaveraged EEG recording; however, they were often obscured in the complexities of the background EEG. Averaging isolates those potentials which are stimulus-related from the activity which occurs randomly with respect to the stimulus. Although there were instances where the background EEG was very noisy, indeed, *e.g.* in the records showing hypsarrhythmia, evoked responses were nevertheless obtained.

The stimuli, sets of 65 dB clicks and bright white and red-orange light flashes, were presented either at a rate of one every 2.5 seconds or at irregular intervals with an average interstimulus interval (ISI) of 4 seconds. The click evoked response was recorded from an electrode at the vertex (Cz) referred to combined mastoids; and the flash evoked response from a midoccipital (Oz)-combined mastoid derivation. Since both age and state of arousal influence the characteristics of evoked responses, the effects of these variables as well as treatment status were analysed.

The relationship of whole blood 5-HI levels to the sensory evoked potentials was examined. The measure for 5-HI used was the one taken closest to the time of the EEG recording, almost always within 1 week. The 5-HI level for each subject was given a rating derived from normal standards: very low (VL), 17–47 ng/ml; low (L), 53–88 ng/ml; low normal (LN), 92–94 ng/ml; normal (N), 97–141 ng/ml; high (H), 196–429 ng/ml.

Auditory Evoked Responses

For statistical analysis, the averaged response to 100 stimuli presented at an ISI of 2.5 seconds while the subject was in EEG sleep stages 2, 3 and 4 was used. Where an individual had several averaged evoked responses obtained under these conditions, the one which showed the highest P_2N_2 amplitude was selected. Fig. 6-5 shows individual responses from both treated and placebo patients. Several prominent peaks are labelled. Peak to trough

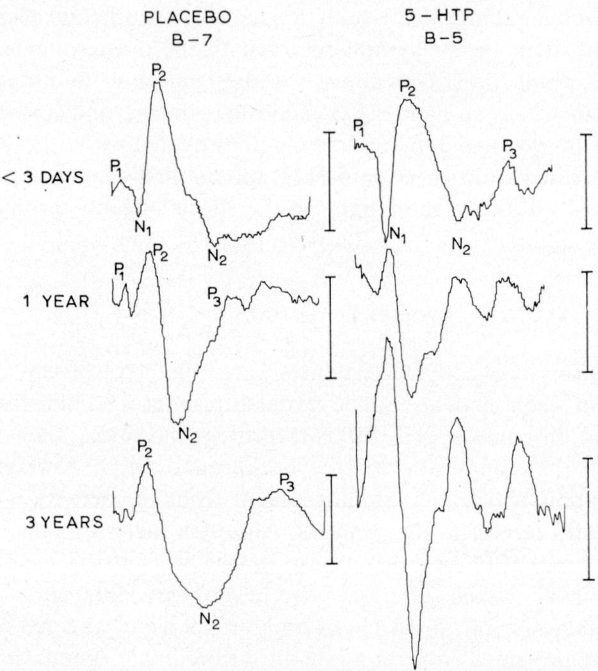

Fig. 6-5. Click evoked responses in Down's syndrome patients. Prominent response components are labelled. The recording is from Cz referred to combined mastoids. An upward deflection denotes positivity of Cz with respect to the reference. The duration of each tracing is 1 second. The stimulus occurred at the beginning of the tracing. Calibration: 25 μV, 1 sec.

amplitude and peak latencies from stimulus onset were measured. No significant differences were found between the groups on D,L- and L-forms of 5-HTP; therefore, these groups were combined for statistical analyses.

The average amplitudes and latencies of several AER components for the 5-HTP treated and placebo patients at five ages are given in Appendices VI-3 and VI-4, and in Figs. 6-6 and 6-7. Comparison between the 5-HTP treated and placebo groups showed significant differences in the prominent

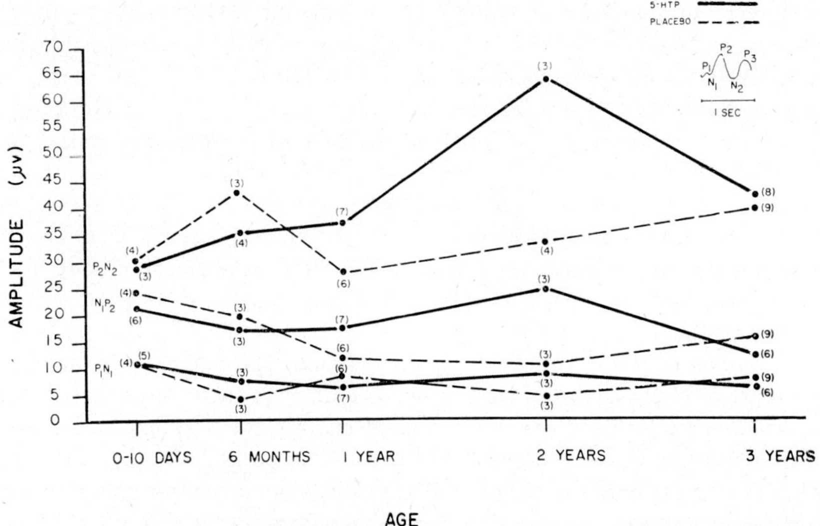

Fig. 6-6. Average amplitudes of AER components in 5-HTP and placebo groups. The numbers in parentheses refer to the number of records used for each point.

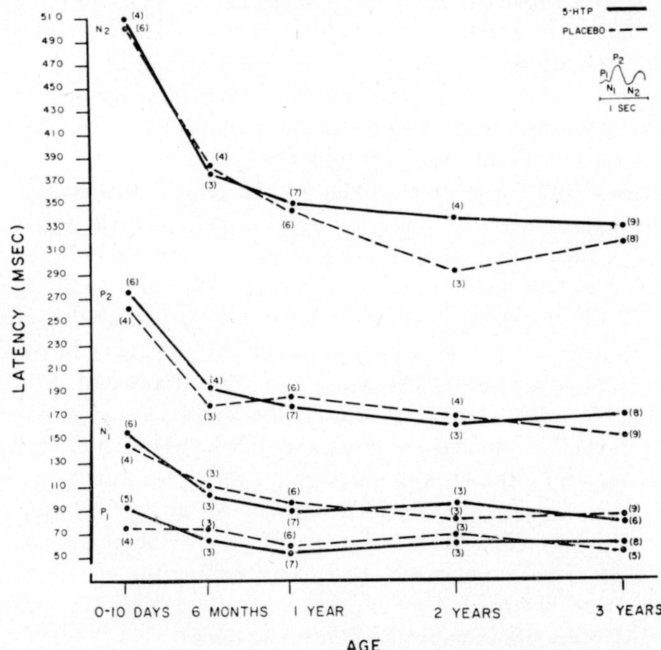

Fig. 6-7. Average latency of AER components in 5-HTP and placebo groups. The numbers in parentheses refer to the number of records used for each point.

early response component, $N_1P_2N_2$. At 1 year, the 5-HTP treated patients had a mean N_1P_2 amplitude of 17.1 μV which was significantly larger than the mean of 10.9 μV for the placebo group ($t=6.210$, $p<0.05$). At 2 years, the mean amplitude of N_1P_2 remained low ($\bar{x}=9.5$ μV) for the placebo group, whereas that of the treated infants reached 25.3 μV ($t=4.646$, $p<0.01$). The amplitude of the P_2N_2 deflection was also greater for the treated infants at 1 and 2 years, but the difference was significant only at 2 years ($t=4.141$, $p<0.01$). Although not statistically significant, the later wave N_2P_3 tended to show a higher amplitude in the 5-HTP group than in the placebo group from 1 to 3 years. By 3 years patients showed similar amplitude for $N_1P_2N_2$ regardless of treatment.

A non-linear relationship between whole blood 5-HI values and amplitude of the wave $N_1P_2N_2$ was evident in the 1-year recordings. It appeared that both extremely low and extremely high levels of 5-HI were associated with small amplitude waves. The N_1P_2 mean amplitude of 8.2 μV obtained from subjects with extremely high blood 5-HI values was not significantly different from the N_1P_2 mean of 11.0 μV from subjects with low or very low 5-HI levels ($t=1.041$, n.s.). The N_1P_2 mean amplitude of 23.4 μV from subjects with 5-HI values in the normal range was significantly higher than either the mean of the low 5-HI subjects ($t=3.978$, $p<0.01$) or the high 5-HI subjects ($t=2.928$, $p<0.01$). The greatest range of 5-HI blood values was observed in the 1-year-old patients with a low of 37 ng/ml and a high of 429 ng/ml. Such extremes were not found in any other age group, therefore comparable analyses at other ages were not possible. Relationships between N_1P_2 amplitude and 5-HI ratings are shown graphically in Fig. 6-8.

The latency of P_2 appeared similar in the 5-HTP treated and placebo groups through 2 years. At 3 years, the mean latency of P_2 for the treated mongols was 165.2 msec (similar to the latency means for both groups at 2 years); however, the placebo group mean of 148.0 msec was significantly earlier ($t=2.187$, $p<0.05$). Significant differences in variability were found between the treated and placebo groups at certain ages; however, there appeared to be no consistent pattern in relation to treatment. For example, N_1P_2 and P_2N_2 amplitudes varied more in the placebo group at 6 months; at 1 year, the 5-HTP group showed more variability in P_1N_1 and N_1P_2 but not in P_2N_2. At 2 years, the groups were similar for P_1N_1 and N_1P_2; however, P_2N_2 was more variable in the treated subjects. No differences were observed at 3 years. Similar inconsistencies in variability were obtained for the latency values of the various components. All t tests comparing group means were for samples with heterogenous variances when there were significant differences in variability between groups (Edwards, 1966).

In addition to the examination of the averaged evoked response to 100 clicks the auditory evoked responses of the 5-HTP and placebo groups were examined with respect to their characteristics during the course of the repetitively presented stimuli. In normal subjects there is usually progressive decline in reponse amplitude as a neutral stimulus is repetitively presented. We have previously shown (Barnet *et al.*, 1971) that this is true of the EEG auditory evoked responses of normal infants at 6 and 12 months but not of Down's syndrome children at the same ages. To determine if 5-HTP affected this habituation or adaption effect, we examined the responses of each patient to the first 25 of a set of 100 stimuli and the last 25 of that set. Amplitude measures were taken from P_2 to N_2 and from N_2 to P_3, and latency measures from stimulus onset to P_2 and P_3 (Fig. 6-5). The frequency of the occurrence

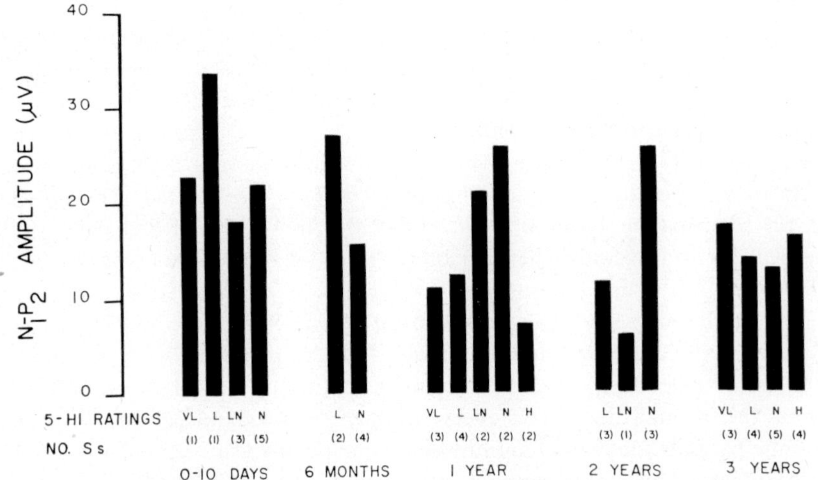

Fig. 6-8. Average amplitudes of N_1P_2 (μV) in the AER of Down's syndrome patients (placebo and 5-HTP treated combined) and ratings of 5-HI blood levels. The range of 5-HI values varies among subgroups with the same ratings.

of response decrement (RD) from the first 25 to the last 25 stimuli in each age and treatment group was calculated.

No significant differences were found between the treated and placebo groups (Fisher exact probability test, Siegel, 1956). No age groups as a whole (all subjects) showed amplitude decrement for either P_2N_2 or N_2P_3; nor did a systematic P_2 or P_3 latency shift occur (Sign test, Siegel, 1956). At 3 years, the recordings of the placebo patients tended to show a decrement in P_2N_2 amplitude ($p = 0.07$, Sign test). No group, placebo or 5-HTP, showed RD for wave N_2P_3 at any age, nor did there seem to be a tendency for the emergence of RD in this late component of the evoked response. This stands in contrast to normal infants who show a high probability for RD for both P_2N_2 and

N_2P_3 components by 6 months of age (Barnet *et al.*, 1971). Thus, 5-HTP did not appear to promote response decrement to repetitive stimulation.

Discussion—AER

A fairly consistent difference over age between the AER of placebo and 5-HTP treated patients was the higher amplitude of the N_1P_2 component in the 5-HTP group. Later response components also tended to be larger. In the mongoloid neonates before treatment, both the N_1P_2 wave and the P_2N_2 wave were often very large. Several other reports have suggested that the amplitudes of some components of the mongoloid AER are abnormally high (Barnet and Lodge, 1967; Straumanis *et al.*, 1970; Ohlrich and Barnet, 1972). Comparable findings were reported for the VER and somatosensory evoked response (Bigum *et al.*, 1970). 5-HTP administration does not appear to cause a decrease in AER amplitude; within the normal range of blood 5-HI the reverse may be true. However, doses resulting in very high blood levels seem to depress the response. This effect may be age related. The patient (A-12) reported by Bazelon *et al.* (1968) did not have AERs which were atypical for mongoloids when her blood levels of 5-HI were extremely high (>1000 ng/ml) at age 3 months.

In a preliminary study (unpublished) of Group A mongoloid patients, we found that the abnormal high amplitude $N_1P_2N_2$ response decreased in size after short periods of 5-HTP administration which raised blood 5-HI to within the normal range. This effect began in the neonatal period and lasted in some patients for several months and sometimes up to 1 year. AER amplitude reduction may have been an acute effect of 5-HTP which disappeared with chronic dosage. However, in the Group A patients, after the neonatal EEG, a procedure was used which may have influenced EEG and evoked response parameters. Immediately prior to the EEG, a femoral venipuncture was performed on some of the patients to obtain blood for chemical studies. This was upsetting, especially to the older infants and may have disrupted their normal sleep patterns. Evoked responses during light sleep are normally lower in amplitude than in deeper sleep. In later studies the venipuncture was done on the day before or after the EEG study.

Although we limited the sample used for evoked response analysis to tracings obtained in EEG sleep stages 2 to 4, the characteristics of these stages differed in the placebo and 5-HTP treated children. Evoked responses are influenced by background EEG characteristics although the nature of the interaction is not always clear. It is possible that the differences noted in the evoked response characteristics of the 5-HTP group may be related to the high voltage, irregular frequencies, paroxysmal abnormalities, and differences in sleep stage organization rather than to differences in the specific sensory system.

Both placebo and 5-HTP treated children in Group B showed developmental changes in their evoked responses which parallel in some respects those of normal children (Barnet and Lodge, 1967; Ohlrich and Barnet, 1972), *i.e.* P_2 and N_2 decreased in latency with age and P_2N_2 and N_2P_3 tended to increase in amplitude. A detailed comparison of mongoloid and normal evoked responses will be the subject of a later report.

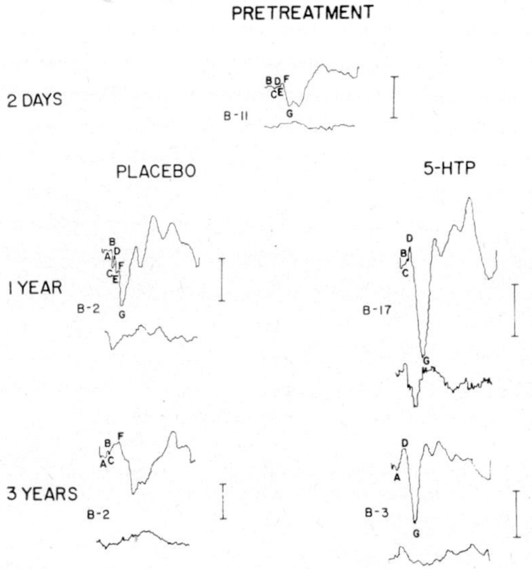

Fig. 6-9. Visual evoked responses summated for 100 flashes presented at a rate of 1/2.5 sec from sleeping Down's syndrome patients, placebo and 5-HTP. The upper tracing of each pair is from a midoccipital (Oz) combined mastoid derivation. The lower tracing is the EOG recorded from supra- and infraorbital electrodes. Prominent components of the VER are labelled. An upward deflection denotes positivity at Oz with respect to the reference. Flash presentation coincided with the beginning of the tracing. Duration of the tracing is 1 second.

Visual Evoked Responses

The data presented were obtained from the averaged responses to 100 white light flashes presented at a rate of 1/2.5 seconds recorded from the midoccipital scalp (Oz). Tracings from infants in sleep stages 1, 2, 3 and 4 were analysed together. Those obtained during wakefulness and stage REM were not included.

Examples of averaged VERs from placebo and 5-HTP treated patients are shown in Fig. 6-9. Because of the varying presence of specific components of the VER in individual records, the peaks were not labelled successively, but rather by latency. Latency criteria were established for specific components by tabulation of peak latencies from all available records. Schematic models of VER patterns may be seen in Fig. 6-10 which shows possible peaks and the manner of labelling after Dustman and Beck (1969). Only 2 recordings presented all components (pattern #3), one placebo patient (B-2) at 1 year and one 5-HTP treated (B-1) at 3 years.

Amplitudes of various deflections are given in Appendix VI-5. The amplitudes presented are limited to measures between adjacent peaks determined by latency criteria with the exception of D-G, which, because of its frequent occurrence was included in the table. Since atypical response forms were sometimes seen, the tabled values are only suggestive of the VER patterns in

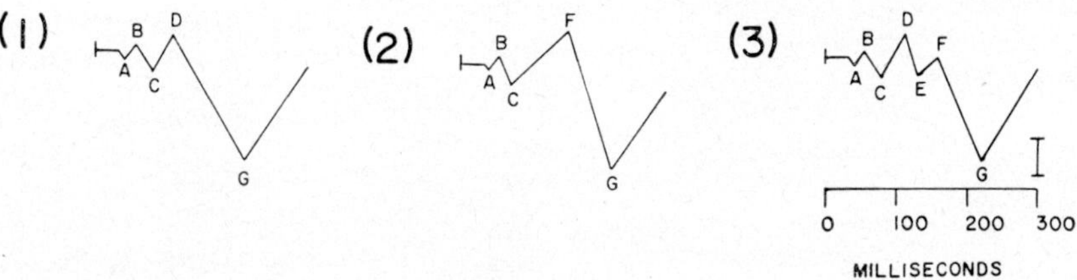

Fig. 6-10. Schematic VER patterns showing possible deflections in the first 300 msec of recordings from Down's syndrome patients.

Down's syndrome infants. Because of the diversity of patterns which was obtained, an amplitude measure from the peak of the most prominent positive wave to the trough of the lowest negative deflection (designated P′–N′) was chosen as an amplitude index for comparative purposes. For example, in Fig. 6-10, P′–N′ in pattern #1 is the D–G deflection; in pattern #2 and #3, it is the F–G deflection. P′–N′ showed a tendency to be larger in the 5-HTP treated groups than in the placebo groups at both 1 year and 3 years. The 5-HTP treated patients had more variable responses at 3 years than the placebo. Although group differences in amplitude values were not significant, the rank order correlation between VER amplitudes and 5-HI values showed a positive relationship at 1 year. Higher 5-HI levels were significantly associated with increased amplitude in treated and placebo patients combined ($r_s = 0.85$, $p < 0.01$). That the correlations were greater or were statistically significant in 5-HTP treated patients but not in the placebo patients, suggests the possibility of a discontinuity between effects obtained with modification of 5-HI levels by 5-HTP administration and effects in untreated patients. Relationships between the amplitude index (P′–N′) and 5-HI values in subjects grouped according to ratings of 5-HI values are shown graphically in Fig. 6-11.

Pretreatment (Neonatal) VERs

In the five recordings from neonates, considerable variability was found in the VER. Four of the five records showed an average latency of 207.1 msec

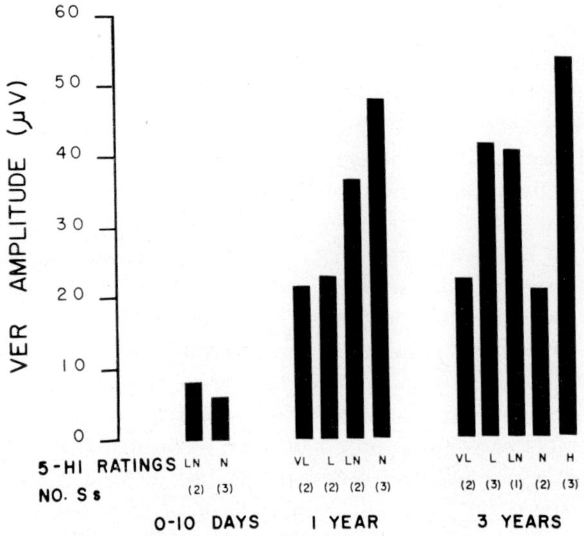

Fig. 6-11. Average VER amplitudes (μV) of Down's syndrome patients (placebo and 5-HTP treated combined) and ratings of 5-HI blood level. The range of 5-HI values varies among subgroups with the same rating.

for the positive peak, F(+). This latency is slightly longer than the reported mean values (184–192 msec) for the prominent positive wave of normal neonates (Ellingson, 1958; Hrbek and Mareš, 1964; Ferriss *et al.*, 1967; Lodge *et al.*, 1969). However, the VER of one neonate (B-13), from an initial small deflection, developed a slow positive wave with the peak at 285 msec. Another record (B-8) did not exhibit a positive wave in the expected range: the largest positive peak was extremely early, at 93 msec, followed by a slow negative peaking at 622 msec. Ellingson (1970) has re-emphasized the characteristic variability of the VERs from normal newborns and his observations would seem to apply to Down's syndrome patients as well.

VERs of 5-HTP and Placebo Groups

Comparisons of peak values (established by latency criteria) did not show statistically significant differences between the placebo and the 5-HTP treated groups. Although differences in variability occurred, these were not consistent. For example, at 3 years of age, the negative deflection E(−) was significantly more variable in the 5-HTP group; however, the later negative G(−) showed greater variability in the placebo records.

Groups are not composed of the same patients at each level, nor are they completely independent. The ranges of 5-HI values vary from one group to another and from one age to another. In the samples from which VERs were analysed, no extreme 5-HI levels occurred in the 1-year-old patients, whereas in the 3-year sample, 3 patients showed high 5-HI. The reverse is true in the samples from which AER data were obtained, *i.e.*, the 1-year group showed very high levels up to 429 n/ml, but the 3-year-old patients did not. The variation in 5-HI levels at different ages and the small samples make the data inadequate to evaluate interactions of age with 5-HI effects.

In the recordings at 1 year, a typical VER pattern was observed (B-17 in Fig. 6-9; #1 in Fig. 6-10). The VERs showed a succession of short latency, distinct but low amplitude waves, A(−), B(+), C(−), which were followed by (D+). D(+) was the most prominent negative peak in 8 of the 9 records. It occurred at an average latency of 112 msec. The VERs of 1-year-old placebo infants tended to have a more complex form than those obtained from 5-HTP infants. Components E(−) and F(+) were observed in 2 out of 3 recordings from placebo patients, but in only 1 of the 6 recordings from 5-HTP treated patients. The largest negative deflection peaked in a fairly narrow latency range, G(−) for all the 1-year-old patients.

In the recordings from both 5-HTP and placebo patients at 3 years, a greater diversity of peak combinations was observed than at 1 year. Although all waves were represented, considerable variation in presence or absence of components in the individual records was noted. Tabulations of peak latencies for individual records are shown in Appendix VI-6. The positive deflection, D(+), found in all 1-year records, was seen in two-thirds of the 3-year records. The later positive, F(+) occurred with greater frequency at 3 years (64%) than at 1 year (33%). Four out of five of the 3-year-old treated infants showed the major positive peak in the D(+) range between 90–114 msec, $\bar{x}=107$ msec, *e.g.* B-3 in Fig. 6-8; in contrast, 4 out of 6 placebo infants had a later major positive peak in the F(+) range, between 138–167 msec, $\bar{x}=156$ msec, *e.g.* B-2 in Fig. 6-9. The trend for an earlier prominent response in 5-HTP treated infants than in placebo infants was not significant (Median test). Most of the patients at 3 years showed a large negative deflection, G(−), with the peak at approximately 215 msec. Considerable variability was observed in the records after this point. For example, in 4 out of the 6 placebo records, the negativity continued after an intervening small positive wave, and the latency of the deepest negative from baseline averaged 363 msec. In contrast, 4 out of 5 treated records showed a large positive wave (22–90 μV) at a similar latency (about 400 msec).

Discussion—VER

Few reports are yet available on the developmental aspects of the VER in mongoloid children. Bigum *et al.* (1970) reported the occipital VERs (O_1 and O_2 derivation) to be more homogenous in 6- to 16-year-old mongoloids than those obtained from normal control subjects. In normal children with an average age of 3 years, Dustman and Beck (1969) found a stable VER with complex early components, and a waveform in the first 250 msec which was comparable to that of adults. In contrast, the 3-year-old mongoloid children in the present study, whether 5-HTP treated or placebo, presented considerable inter-individual inconsistency of both early and late response components.

On comparison of our data on 3-year-old children with that obtained for older mongoloids (Bigum *et al.*, 1970) and for normal children with a mean age of 3 years (Dustman and Beck, 1969), the latencies of the early components A(−), B(+) and C(−) agree fairly well. However, the longer latency components in our 3-year-old sample occur somewhat earlier; the F(+) mean was 156 msec for placebo subjects and 150 msec for 5-HTP treated subjects. The reported F(+) value for normal 3-year-olds was 191.6 msec (Dustman and Beck, 1969); and for older mongoloids, 182.5 msec (Bigum *et al.*, 1970). The short F(+) latency in our sample may be attributable to differences in experimental conditions, *e.g.* illumination or subjects' levels of consciousness.

Other investigators have described larger VER amplitudes in mongoloids than in normals (Bigum *et al.*, 1970; Straumanis *et al.*, 1970). The deviant or markedly irregular VER waveforms observed in the present study make meaningful comparisons with published data difficult. The variability of the waveform observed suggests disturbances in maturation and function of the developing sensory systems. The fact that VERs of mongoloids were modified by changes in 5-HI blood levels may imply that cerebral serotonin metabolism plays a significant role in the development of cortical visual information processing.

Summary

Serial electroencephalograms and sensory evoked response recordings were obtained on 19 group B (double blind) patients with Down's syndrome.

Recordings were performed in the neonatal period before 5-HTP administration was begun, and at 6 months, 1, 2 and 3 years of age. They were performed during unsedated daytime sleep.

There was a statistically significant increase in EEG abnormality in the 5-HTP treated group over the placebo group and one L-5-HTP treated patient developed infantile spasms. The patient with clinical seizures had a hypsarrhythmic EEG. Her seizures persisted despite discontinuation of 5-HTP. Another patient on the D,L-form of 5-HTP showed a hypsarrhythmic EEG at 2 years but had no seizures. Other EEG abnormalities in the 5-HTP patients included paroxysmal slowing, spikes, polyspikes and slow waves.

Organization of EEG sleep stages appeared to be related to treatment. Stage REM was reduced at 6 and 12 months in the 5-HTP patients in comparison to the placebo patients. By 2 years of age, both treated and placebo subjects showed little REM sleep. The amount of high voltage slow wave sleep (stages 3 and 4) increased with age in both groups; however, the 5-HTP groups by 1 year of age had more high voltage slow wave sleep than the placebo patients. Slow wave sleep in patients with high whole blood 5-HI was often of extremely high voltage and was mixed with irregular fast activity. In all groups, indeterminate sleep occupied a considerable portion of the total sleep time.

Auditory and visual evoked potentials were affected by treatment status. However, no response patterns emerged which were consistently associated with treatment. VERs and AERs in both placebo and 5-HTP treated patients with Down's syndrome were abnormal and although 5-HTP changed evoked response patterns, it did not 'normalize' them.

Short latency components of the AER showed higher amplitude in the 5-HTP treated patients than in the placebo patients. VERs exhibited extreme variability in both groups, but the VERs of the 5-HTP group tended to be simpler in form than those of the placebo patients and to have a shorter latency for the major response component. Neither placebo nor 5-HTP treated subjects showed decrement in AER amplitude during repetitive presentation of the stimulus. The relationship between blood 5-HI levels and evoked responses is not clear-cut. However, the high AER amplitude associated with elevation of blood 5-HI levels to normal range in treated patients, the non-linear relationships between 5-HI and AER amplitude at 1 year, and the positive correlation between VER amplitude and 5-HI level at 1 year, suggest some regulatory mechanism relating changes in cerebral serotonin metabolism to modification of cortical potentials.

REFERENCES

ANDERS, T., ENDE, R. and PARMELEE, A. (eds.) (1971) A manual of standardized terminology techniques and criteria for scoring of states of sleep and wakefulness in newborn infants. UCLA Brain Information Service, Los Angeles, California.

ASERINSKY, E. and KLEITMAN, N. (1955) A motility cycle in sleeping infants as manifested by ocular and gross bodily activity. *J. Appl. Physiol.* **8**, 11.

BARNET, A. B. and Lodge, A. (1967) Click evoked EEG responses in normal and developmentally retarded infants. *Nature* (*Lond.*) **214**, 252.

BARNET, A. B., OHLRICH, E. S. and SHANKS, B. L. (1971) EEG evoked responses to repetitive stimulation in normal and Down's syndrome infants. *Dev. Med. Child Neurol.* **13**, 321.

BAZELON, M., BARNET, A. B., LODGE, A. and SHELBURNE, S. A. (1968) The effect of high doses of 5-hydroxytryptophan on a patient with trisomy 21. *Brain Res.* **11**, 397.

BIGUM, H. B., DUSTMAN, R. E. and BECK, E. C. (1970) Visual and somato-sensory evoked responses from mongoloid and normal children. *Electroenceph. clin. Neurophysiol.* **28**, 576.

DREYFUS-BRISAC, C. (1964) The electroencephalogram of the premature infant and full term newborn: normal and abnormal development of waking and sleeping patterns. In: Kellaway, P. and Petersen, I. (eds.), Neurological and electroencephalographic correlative studies in infancy, Grune and Stratton, New York, p. 186.

DREYFUS-BRISAC, C. and MONOD, N. (1970) Sleeping behavior in abnormal newborn infants. *Neuropaediatrie*, **3**, 354.

DUSTMAN, R. E. and BECK, E. C. (1969) The effects of maturation and aging on the waveform of visually evoked potentials. *Electroenceph. clin. Neurophysiol.* **26**, 2.

EDWARDS, A. L. (1966) Experimental design in psychological research (rev. ed.), Holt Rinehart and Winston, New York.

ELLINGSON, R. J. (1958) Electroencephalograms of normal, full-term newborns immediately after birth with observations on arousal and visual evoked responses. *Electroenceph. clin. Neurophysiol.* **10**, 31.

ELLINGSON, R. J. (1970) Variability of visual evoked responses in the human newborn. *Electroenceph. clin. Neurophysiol.* **29**, 10.

ELLINGSON, R. J. (1973) EEG disorders associated with chromosome anomalies. In: International handbook of EEG and clinical neurophysiology, A. Remond (ed.), Vol. 13, Theme 32. Hereditary, Congenital and Perinatal Diseases, in press.

ELLINGSON, R. J., MENOLASCINO, F. J. and EISEN, J. D. (1970) Clinical-EEG relationships in mongoloids confirmed by karyotype. *Am. J. Ment. Defic.* **74**, 645.

FEINBERG, I., BRAUN, M. and SHULMAN, E. (1969) EEG sleep patterns in mental retardation. *Electroenceph. clin. Neurophysiol.* **27**, 128.

FERRISS, G. S., DAVIS, G. D., DORSEN, M.MCF. and HACKETT, E. R. (1967) Changes in latency and form of the photically induced average evoked response in human infants. *Electroenceph. clin. Neurophysiol.* **22**, 305.

GIBBS, E. L. and GIBBS, F. A. (1962) Extreme spindles: correlation of electroencephalographic sleep patterns with mental retardation. *Science*, **138**, 1106.

GIBBS, F. A. and GIBBS, E. L. (1952) Atlas of electroencephalography, Vol. 2, Addison-Wesley Publishing Co., London, p. 92.

GOLDIE, L., CURTIS, J. A. H., SVENDSEN, U. and ROBERTON, N. R. C. (1968) Abnormal sleep rhythms in mongol babies. *Lancet*, **1**, 229.

HRBEK, A. and MAREŠ, P. (1964) Cortical evoked responses to visual stimulation in full-term and premature newborns. *Electroenceph. clin. Neurophysiol.* **16**, 575.

JASPER, H. H. (1958) The ten-twenty electrode system of the International Federation. *Electroenceph. clin. Neurophysiol.* **10**, 371.

JOUVET, M. (1969) Biogenic amines and the states of sleep. *Science*, **163**, 32.

KELLAWAY, P. (1952) The development of sleep spindles and of arousal patterns in infants and their characteristics in normal and certain abnormal states. *Electroenceph. clin. Neurophysiol.* **4**, 369.

LEE, J. C. M., ORNITZ, E. M., TANGUAY, P. E. and RITVO, E. R. (1969) Sleep EEG patterns in a case of Down's syndrome—before and after 5-HTP. *Electroenceph. clin. Neurophysiol.* **27**, 686 (abstract).

LODGE, A., ARMINGTON, J. C., BARNET, A. B., SHANKS, B. L. and NEWCOMB, C. (1969) Newborn infants' electroretinograms and evoked electroencephalographic responses to orange and white light. *Child Dev.* **40**, 267.

MANDELL, A. J. and MANDELL, M. P. (1965) Biochemical aspects of rapid eye movement sleep. *Am. J. Psychiat.* **122**, 391.

MURRAY, B. and CAMPBELL, D. (1971) Sleep states in the newborn. Influence of sound. *Neuropaediatrie*, **21**, 335.

OHLRICH, E. S. and BARNET, A. B. (1972) Auditory evoked responses during the first year of life. *Electroenceph. clin. Neurophysiol.* **32**, 161.

PARMELEE, A. H., AKIYAMA, Y., SCHULTZ, M. A., WENNER, W. H., SCHULTE, F. J. and STERN, E. (1967) The electroencephalogram in active and quiet sleep in infants, Int'l. Symp. Clin. EEG in Childhood, Goteborg, Sweden, August.

PETRE-QUADENS, O. and DEGREEF, A. (1971) Effects of 5-HTP on sleep in mongol children. A preliminary report. *J. Neurol. Sci.* **13**, 115.

PRECHTL, H. F. R., WEINMANN, H. and AKIYAMA, Y. (1969) Organization of physiological parameters in normal and neurologically abnormal infants. *Neuropaediatrie*, **1**, 101.

RECHTSCHAFFEN, A. and KALES, A. (eds.) (1968) A manual of standardized terminology, techniques and scoring system for sleep stages of human subjects, USPHS, Washington, D.C.

ROFFWARG, H. P., MUZIO, J. N. and DEMENT, W. C. (1966) Ontogenetic development of the human sleep-dream cycle. *Science*, **152**, 604.

ROHDE, J. A., KOOI, K. A. and RICHEY, E. T. (1969) Sleep spindles, mental retardation and epilepsy. *Electroenceph. clin. Neurophysiol.* **26**, 12.

SIEGEL, S. (1956) Nonparametric statistics for the behavioural sciences, McGraw-Hill, New York.

STRAUMANI Jr, J. J., SHAGASS, C. and OVERTON, D. A. (1970) Evoked responses in Down's syndrome of young adults. *Electroenceph. clin. Neurophysiol.* **29**, 321 (abstract).

WYATT, R. J., ZARCONE, V., ENGELMAN, K., DEMENT, W. C., SNYDER, F. and SJOERDSMA, A. (1971) Effects of 5-hydroxyptophan on the sleep of normal human subjects. *Electroenceph. clin. Neurophysiol.* **30**, 505.

CHAPTER 7

Acute and Chronic Side Effects of 5-Hydroxytryptophan

MARY COLEMAN and ANN BARNET

Most patients with Down's syndrome tolerate 5-HTP with surprisingly few clinical side effects. To date there appears to be one major toxicity—the development of the infantile spasms syndrome with or without EEG abnormalities in 14% of the young patients receiving 5-HTP on a chronic basis.

In our clinic, 62 patients received 5-HTP, often starting in the neonatal period. No clinic patients still receive 5-HTP. The acute and chronic toxicities we observed with 5-HTP in Down's syndrome patients are seen in Table 7-1 and Appendix VII-1.

TABLE 7-1

Side effects of 5-hydroxytryptophan in 62 patients with Down's syndrome

Side effect	Percentage of total
ACUTE	
1. *Gastrointestinal symptoms*	
Diarrhea	43
Gastric pain	1.6
2. *CNS symptoms*	
Hyperactivity	21
Opisthotonos	10
Hyperacousis	6
3. *Vascular symptoms*	
Hypertension	8
Acrocyanosis	1.6
CHRONIC	
1. *Convulsive phenomena*	17
Infantile spasms type	(14)
2. *?Glaucoma*	? 3
3. *?Ectodermal lesions*	? 3

When one follows 62 children from the neonatal period up through 2–5 years, obviously many changes and illnesses occur spontaneously; that is, unrelated to the administration of 5-HTP. Table 7-1 lists side effects which we have reason to suspect were related to the amino acid administration, either because of unusual frequency of occurrence in this patient group or because of premature termination of the side effect by lowering the 5-HTP dosage in individual patients.

The most common side effect of 5-HTP in this patient group was diarrhea. Almost all patients had diarrhea, often with fever and other symptoms at some time during this study. In 27 of the patients, at one or more times during these early years, the diarrhea occurred without other clinical symtoms, usually when the dosage was raised quickly; it could be stopped by lowering the 5-HTP dosage. In many of these patients, this pattern repeated several times. The majority of cases occurred during the first year of life but there were several 2-year-olds (? particularly susceptible) who had to be virtually taken off 5-HTP to control a chronic diarrhea. These children had extensive negative workups for the usual organic etiologies of diarrhea. One patient, twin D-22, had gastrointestinal symptoms so severely that she was thought to have Hirschsprung's disease; however, the rectal biopsy was normal. Other patients in the study group had constipation (the usual pattern seen in Down's syndrome) often alternating with normal bowel movements. A few patients were essentially normal throughout the study.

The other troublesome gastrointestinal side effect of 5-HTP was acute gastric burning pain occurring 20 minutes after ingestion of 5-HTP dosages in a 4-year-old who was advanced enough to describe the symptoms to us. In this patient, administration of 5-HTP with food greatly relieved the symptom. When she was tapered off 5-HTP, the symptoms stopped except for "burping with a rotten egg odor" after some weeks.

There were a number of patients, usually young infants, who had spitting up or vomiting after feeding. Lowering the dose of 5-HTP had no effect on vomiting or spitting up except in one questionable case. If this is a side effect of 5-HTP, it must be a rare one.

Patient D-2 was an interesting child with trisomy 21 who had a paradoxical tonus response to 5-HTP (she became even more hypotonic) and also a paradoxical gastrointestinal response. She developed diarrhea each time her dosage was *lowered.* Apparently she is the exception to a general rule.

The second most prominant side effect of 5-HTP seen in this group of patients was abnormally increased activity. Five of the 62 patients in this study were very active as infants and somewhat hyperactive as 2- and 3-year-olds. They were so classified on almost every clinic visit, regardless of 5-HTP dose. We assumed that they were probably part of that roughly 10% of Down's syndrome patients who are spontaneously hyperactive and did not classify their behavior as a 5-HTP side effect. However, in 13 patients in this study (21%) a type of overtly abnormal overactivity was seen that could be stopped by reduction of the dose of 5-HTP. Eight of the patients were less than 4 months of age, a period when spontaneously better activity and turning over is sometimes seen in totally untreated Down's syndrome patients (*i.e.* placebo patients B-15 and B-13). In one case (A-5 listed in the side effects table, a 13-month-old trisomy), the markedly increased activity level followed an accidental double dose of 5-HTP. Irritability and restlessness also have been reported in patients if the dose is increased too quickly or, in retrospect, the patient was overdosed. Irritability has also been reported in patients for the first 6 weeks they are tapered off 5-HTP.

True insomnia has not been seen in our patient group but a probable decrease in total hours of sleep has occurred in 5 patients. Several patients were reported to be 'always poor sleepers' with 3-year-old trisomics having only 8 out of 24 hours of sleep on some nights. Patient B-1, who has a normal twin brother, sleeps 1 hour less than her sibling at night and 30 minutes less at nap time. One patient was reported restless at night, sometimes 'crying' in her sleep. Untreated patients generally sleep very well, too well, but there are occasional sleeping problems in Down's syndrome. So it is hard to be sure that 5-HTP *per se* was responsible for any significant loss of sleep in our patient group. Only 1 of the patients with possible decreased sleep time (D-12) also was overactive in daylight hours.

Opisthotonos was seen in the first 4 months of life in 6 patients. Perhaps by chance, 2/3 of these patients were in the double blind study (Group B), a group where dosage regulation was done solely by clinical examination without the benefit of the blood chemistries. However, in only 1 of the patients was the total 5-HI really elevated at the time opisthotonos was documented. In some of the older patients (9–16 months) arching of the back with crying or when held occurred but we did not consider this type of voluntary posturing to be opisthotonos.

The report of hyperacousis by 4 parents was an interesting finding. The earliest report at 6 weeks of age was in the double blind patient (B-1) who had a normal twin brother for comparison. At that time, it was considered a possible natural variation between 2 individuals. However, later reports in other patients emphasized difficulty with loud noises (patients turning down loud televisions, hating noisy crowds). In one patient (D-1) it was a continuing problem first recorded at 14 months of age and persisting intermittently up to 3 years of age.

As mentioned earlier in Chapter 2, ataxia was a prominent symptom when 5-HTP was withdrawn too quickly. It was also reported as a side effect of overdosage in one patient (D-9) listed in Appendix II-1 during the period of overtreatment, prior to withdrawal of 5-HTP. However, this was the only report of ataxia with high levels and was difficult to evaluate because the 2-year-old patient had just learned to walk and variations in balancing from day to day are seen in normal children. The only other hint of ataxia in our records is a report of a 1-year-old trisomy who became 'wobbly and glassy-eyed' when the dose was raised too quickly.

Hypertension was seen in the very young patients and mostly in the initial group of patients when we were first learning how rapidly to increase dosage rates. Like many acute side effects, it was the rate of increment, rather than

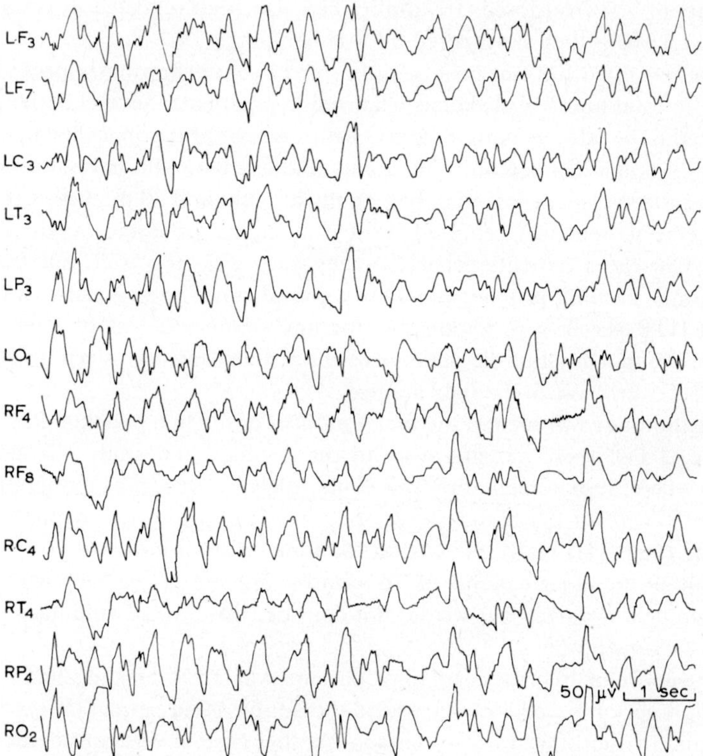

Fig. 7-1. In this patient with trisomy 21, the high voltage disorganized mixture of spikes and slow waves persisted after discontinuing 5-HTP. Tracing was taken one week later. (Reprinted with permission of *Neurology*.)

the absolute value, of the dose of 5-HTP that appeared to produce the toxic effect. One of the later patients (D-25) with hypertension was a patient with multiple other medical problems (just out of a respirator for apnea secondary to bilateral pneumonia).

Acrocyanosis of the extremities was seen in only 1 patient without cardiac disease—Patient A-12, whose parents inadvertently raised the dosage of 5-HTP to extremely high levels (Bazelon *et al.*, 1968). The acrocyanosis gradually cleared on this child but reappeared again at the age of 12–14 months when relatively high dosages of 5-HTP again were given (5-HI, 386 ng/ml, 322 ng/ml). At 16 months of age, on lower 5-HTP dosages (5HI, 177 ng/ml), the acrocyanosis finally disappeared. Extremity pain was reported in association with 5-HTP dosages by 2 older patients; in 1 case, the feet hurt, in another, the legs hurt and had cramping, which stopped when 5-HTP was discontinued.

The major chronic toxicity of 5-HTP in Down's syndrome patients were convulsive phenomena, see Appendix VII-2.

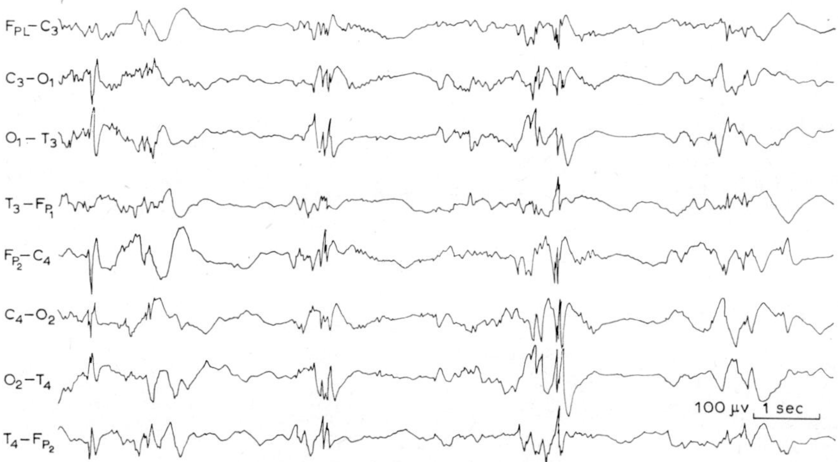

Fig. 7-2. Electrodecremental intervals intersperse the high voltage spiking and slow waves in this trisomy 21 patient still receiving 5-HTP. (Reprinted with permission of *Neurology*.)

The patients with clinical seizures fell into two distinct groups. In the first group of 4 patients, the EEG was abnormal and clinical seizures were not stopped by lowering or terminating 5-HTP. In these patients the EEG abnormalities were those of hypsarrhythmia with very high voltage disorganized blends of spikes and slow waves (Fig. 7-1), sometimes interspersed with electrodecremental intervals (Fig. 7-2). The EEG abnormalities in 1 patient, D-15, were reversed by a course of ACTH (Fig. 7-3), although the patient was left severely hypotonic afterwards and is now institutionalized.

The second group of patients with clinically overt seizure phenomena had normal EEG's and seizures reversed by lowering the dose of 5-HTP. Of these 7 patients with normal EEGs, 5 had an onset after 9 months of age with seizures of the infantile spasm type resembling those of the other patient group with EEG abnormalities. However, 2 patients with normal EEG's had brief seizure phenomena, at 4 days and 6 weeks of life (Appendix VII-2). In patient B-10, several episodes of rhythmic twitching of the left arm and leg were noted 24 hours after oral L-5-HTP was begun. The solution (unknown at

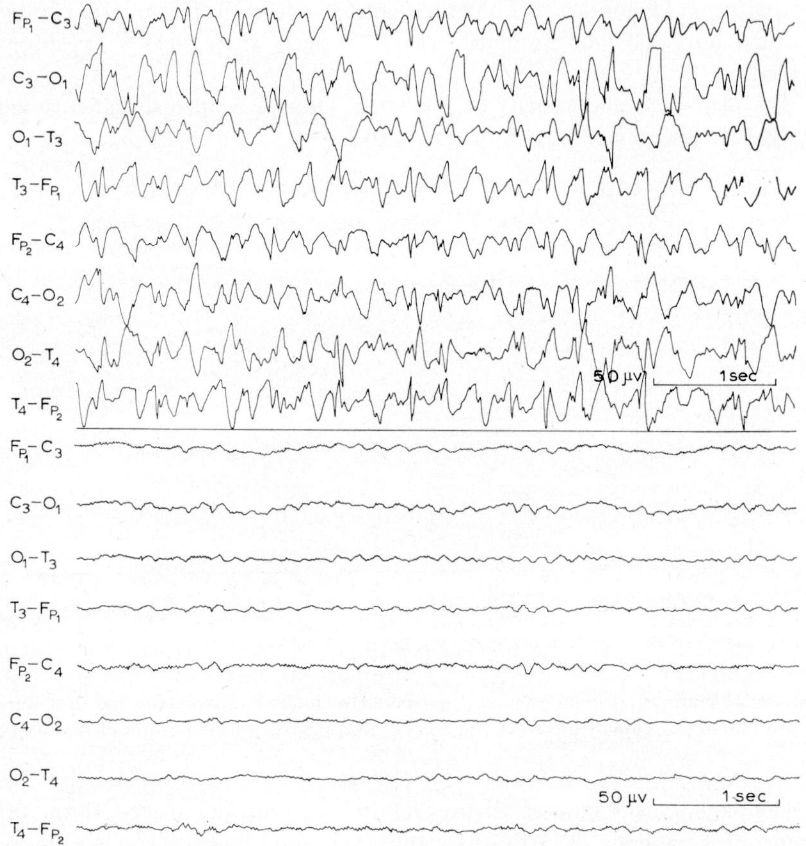

Fig. 7-3. Disorganized spikes and slow waves (top) persisted in the EEG of this trisomy 21 patient after discontinuation of 5-HTP. A 'normal' EEG (bottom) and the end of clinical seizures in the patient occurred after a 3-month course of ACTH.

that time since the patient was on the double blind study) was discontinued, the twitching stopped, a seizure work-up was completely negative and L-5-HTP was begun again at 9 days of age without further incidents. In patient D-14, there were 3 episodes (at 6 weeks, 19 months and 22 months of age) of brief clonic movements seen only by the parents.

The clinical seizure patterns seen in most of these children closely resembled the infantile spasm syndrome. The spasms of the rest of the patients tended to fall into salaam or flexion spasm pattern repeated a number of times to form a rapid series. Extensor spasms and nodding spasms (flexion of the neck) were less often seen or described by parents. Staring eye movement and eyelid blinking was seen in all patients. In patients C-32, D-4, and F-2 a

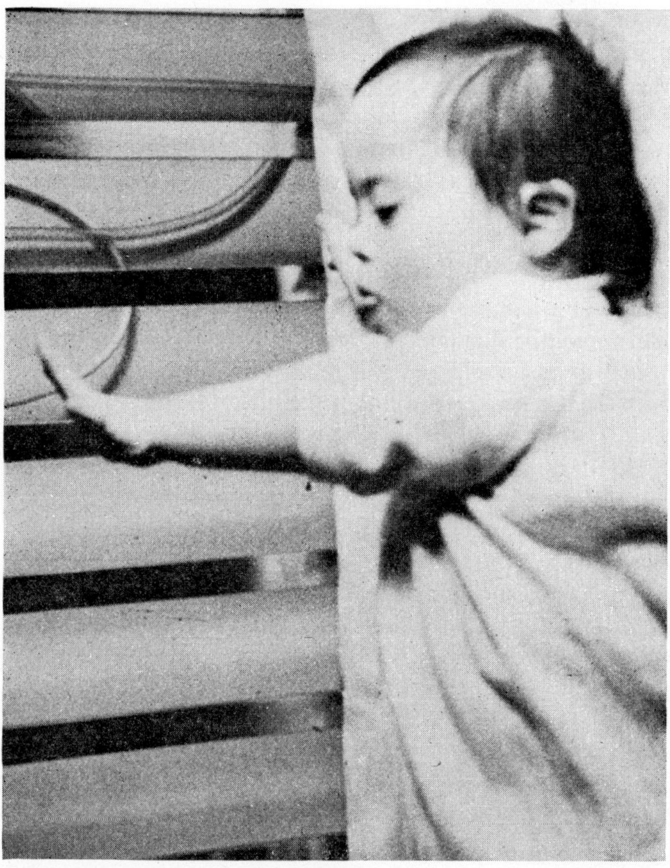

Fig. 7-4. Part of the infantile spasm pattern in these patients was a stereotyped posture of the baby, apparently 'staring' with miotic pupils at one rigidly outstretched hand.

striking stereotyped posture of the baby, apparently staring with miotic pupils at one rigidly held outstretched hand, appeared to be part of the seizure pattern in addition to the more classical signs (Fig. 7.4). In the 4 irreversible cases, in retrospect, all mothers noted social withdrawal and inability to get the child to relate interpersonally, usually more than 1 month prior to the actual onset of seizures. In these irreversible cases, cessation of 5-HTP improved the frequency and severity of the seizures, but did not totally halt the convulsive pattern at first. In patient B-IS-19, the seizures eventually stopped after 14 months, and reoccurred when the child was replaced on the unknown solution (which turned out to be L-5-HTP) at the parents' insistence in an effort to overcome hypotonia. Eventually the patient tolerated low dosages of 5-HTP without seizures. This child remains with her family. Patient D-7 died after the parents had moved 2000 miles away; patient C-32, also an out of town patient, is institutionalized and the parents refused a course of ACTH. The reversal of the clinical seizures and the EEG abnormalities by ACTH in patient D-15 helps confirm that this patient had the infantile spasm syndrome either spontaneously or 5-HTP induced.

Review of clinical histories showed that several of the patients appeared to be especially sensitive to 5-HTP. They showed signs of 5-HTP toxicity (diarrhea, overactivity, hypertonus) at lower doses and with lower 5-HI values (than most patients) when they first began receiving the amino acid and well before the clinical seizure pattern developed. This suggests that high risk patients could be identified early and carefully followed in an effort to abort the development of the syndrome. The seizure phenomenon began usually with the patient grading close to or normal on the neurological tonus examination in those patients who were started in the newborn period on 5-HTP. Also, in at least one instance, moderate hypotonia was still present in a patient who developed seizures after being started on 5-HTP later than the newborn period. No patient was hypertonic at the time seizures began.

There were a sizable number of children in this study who had EEG abnormalities without clinical seizure phenomena. One patient had a frankly hypsarrhythmic pattern at 2 years of age and a grossly abnormal EEG at 3 years (B-3), but *no seizures*, social withdrawal or evidence of mental or physical deterioration.

In the double blind placebo group, 2 subjects (22%) showed EEG abnormalities and 1 subject (11%) a borderline abnormal record. Both were seizure-free. The percentage of EEG abnormalities in the placebo group was within the range reported in the literature. Two D,L-5-HTP treated children in the double blind study (40%) had abnormal EEGs. In the L-5-HTP groups abnormal EEGs were obtained from 60% (3 patients) and a borderline abnormal record from a fourth patient. Thus, although 11 of the 19 children (58%) in the double blind study had EEG abnormalities, only 1 in this group (B-IS-19) had clinical seizures. The significantly higher incidence of EEG abnormality in the 5-HTP group compared with the placebo group lends weight to our assumption that the drug was a causal factor both in symptomatic and asymptomatic EEG abnormalities. For further discussion of EEG changes during 5-HTP administration, see Chapter 6.

Although the infantile spasm syndrome is the only chronic toxicity that appears to be definitely associated with 5-HTP, several other possible toxicities are included here. Two trisomy 21 patients (A-3 and D-22) on the 5-HTP regimen developed glaucoma. The reports of their ophthalmologists are seen in Appendix VII-3. In both cases it was unclear whether it was infantile glaucoma with dystrophy of the cornea secondary to factors after birth, such as 5-HTP, or whether prenatal structural factors may have contributed to the disease process.

However, 1 of the patients who developed glaucoma (Patient D-22) was a twin who received 5-HTP while her identical twin sister (Patient E-19) received only B6. Both twins were started on their respective therapies at the age of 2 days. Although the first twin who received 5-HTP had bilateral disease with marked breakage in the cornea, the second twin (E-19) receiving B6 never had the slightest evidence of change in ocular pressures. Twin D-22 with the infantile glaucoma also had hematological abnormalities, and a severe gastrointestinal disease process at 5 weeks of age thought to be Hirschsprung's disease but without X-ray evidence of that disease. Of the 2 patients with glaucoma, twin D-22's ocular problems were first noted 15 weeks after initiation of 5-HTP therapy; patient A-3's symptoms were definitely diagnosed 3 years after the amino acid was begun. Discontinuation of D,L-5-HTP had no effect on the clinical course of patient A-3; a repeat goniotomy had to be performed—months after discontinuing the amino acid.

A number of ectodermal lesions were seen in the patients through the years. Two of them were troubling and may be side effects of 5-HTP (Appendix VII-4.) One patient (D-8) developed a macular skin rash and alopecia areata several years after beginning 5-HTP and B6 in the neonatal period (see Fig. 7-5). She began losing her hair at 23½ months of age when she was on 5-HTP. A scalp biopsy disclosed mononuclear cell infiltration surrounding the residual hair follicles, confirming the clinical impression of alopecia areata (see Appendix VII-4). None of the disease entities correlated with alopecia areata (thyroid disease, diabetes, migraine headaches, psychiatric disease (Muller and Winkelmann, 1963)) could be confirmed in this patient. It is noted for the record that the patient's 43-year-old mother was hospitalized with severe depression (thought to be a reactive depression to the child's diagnosis) when the child was between 2 and 4 months of age. However, since that time she appeared to be quite well accepted by her parents and was one of the patients doing relatively well on the 2-year examination when the hair loss was first documented. Six months after the hair loss began, 5-HTP was stopped and 1 month later regrowth began. However, it is difficult to be sure that discontinuance of 5-HTP was a crucial factor in regrowth since the regrowth after a few months occurs in the natural history of alopecia areata.

A granulomatous skin rash (Appendix VII-4) developed on the face of patient A-5 and persisted until she was tapered off 5-HTP. The pathologists raised the possibility of a granulomatous drug eruption. In the absence of other cases, however, the relationship to 5-HTP remains unsettled. The rash improved with cortisone ointment.

Other ectodermal lesions seen in patients in the study group were less definitely connected to the amino acid. Diaper rash accompanying the diarrhea of toxic levels of 5-HTP was reported. Heat rash, cracked lips, dermatitis related to sun exposure, and petechiae after trauma are seen in untreated patients with Down's syndrome as well as in our patient group.

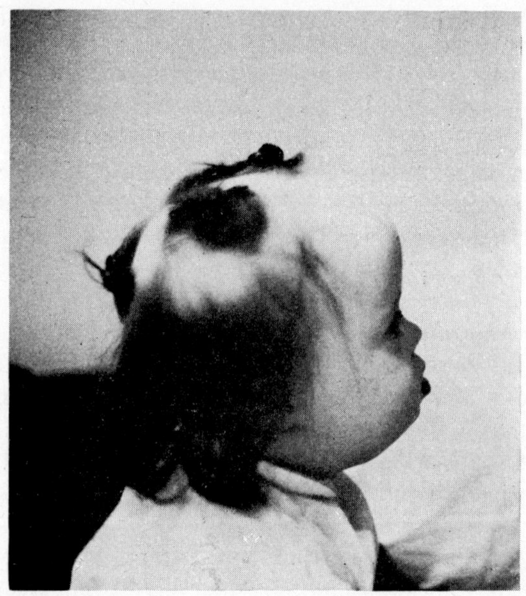

Fig. 7-5. Alopecia areata occurred in 2-year-old trisomy female receiving D,L-5-HTP and vitamin B6. None of the other 61 patients who received 5-HTP had a similar hair loss.

Several patients, both placebo and 5-HTP, had dental cavities by 4 years of age. In the case of patient B-7, who refused milk after 5 months of age, multiple cavities and dark staining of the teeth were reported.

Laboratory abnormalities accompanying long-term 5-HTP therapy were surprisingly few. Anemia was present in 2 patients in the study. One child (B-9) had a chronic anemia which showed very minimal improvement with the addition of vitamin B6 at 3 years of age and tapering off the 5-HTP. Elaborate work-up showed no known etiology of the anemia. Twin D-22 also had hemolytic anemia, leukocytosis and thrombocytopenia at 5 weeks of age during the crisis when she was removed from 5-HTP. Hematological abnormalities never reoccurred. SGOT and SGPT levels elevated in 20% of all patients receiving 5-HTP, but also in 20% of the placebo group—confirming an earlier report by Saxl (1968). The only abnormality noted on urinalysis was uric acid crystals in one 3-year-old receiving the D,L-form. Hyperuricemia is known to occur spontaneously in Down's syndrome.

Discussion

There does not seem to be a close correlation between the dosage of 5-HTP or whole blood 5-HI level and the development of toxic symptoms of 5-HTP administration. Some individuals seemed to be especially sensitive to the effect of the amino acid; often they were patients who had neurological tonus examinations better than usually seen in their age group in Down's syndrome.

In general, acute side effects were most likely to occur in three circumstances: (1) when initially starting, (2) when the rate of increase of 5-HTP is excessive, or (3) when a child has a fever or other systemic illness. Many of the acute (and possibly some of the chronic toxicities) of 5-HTP tended to be due to the *rate of increment* of the amino acid rather than the absolute level achieved at any particular time. If done gradually, 5-HTP can be pushed to high levels in some patients with Down's syndrome with surprisingly few toxic side effects (Bazelon *et al.*, 1968).

Since there is experimental evidence that 5-HTP stimulates intestinal motility (Haverback and Davidson, 1958) it was no surprise that diarrhea was a major acute side effect of the amino acid. Two disease entities with diarrhea (the carcinoid syndrome and non-tropical sprue) are reported to have elevated 5-HIAA levels. 5-HTP given intravenously to adult schizophrenics also receiving a monoamine oxidase inhibitor produced acute nausea and vomiting rather than diarrhea (Pollin, 1961).

Partington *et al.* (1971) gave D,L-5-HTP to 6 Down's syndrome children and 6 other retardates for 1 week intervals. Presumably they began without any tapering or working up to the prescribed dose. On patients started on 2 mg/kg dosages given 3 times each day (total—6 mg/kg/day), they reported that 1 child developed a fever with vomiting for 24 hours and another a fever with blood in the stool. In contrast to 5-HTP, which induced diarrhea in our patient group, pCPA (a tryptophan hydroxylase inhibitor that lowers 5-HT) stops diarrhea in patients with the carcinoid syndrome. It is interesting to contrast the effect of 5-HTP with L-Dopa in the gastrointestinal tract. In patients with Parkinson's disease (Cotzias, 1968) and dystonia (Coleman, 1970), too rapid an increase of L-Dopa induced vomiting; of 5-HTP in this monograph, diarrhea. In both sets of patients with both amino acids, gastrointestinal symptoms were most prominent in early stages of administration and seen only rarely after the first months.

Hyperacousis, hyperactivity and hypertension have all been reported before in many 5-HTP experiments in both animals and humans (Erspamer, 1966; Garattini and Valzelli, 1965). Partington *et al.* (1971) began Down's syndrome patients and other retardates on 6 mg/kg/day of the D,L-form of 5-HTP and reported hypertension in 18% of their series during the first 3 days. The blood pressures returned to non-treatment levels by the 4th day.

The main side effect that caused much concern during this study was the development of seizures in 17% of the patients. With 2 exceptions, the seizures were of the type usually classified as the 'infantile spasm syndrome'. This syndrome (also known as infantile myoclonic jerks or minor motor epilepsy of infancy) is thought to be the result of seizure activity of an immature nervous system usually presenting between 2 and 36 months of age and with many different etiologies, including both focal and diffuse brain disease. It is one of the seizure syndromes often associated with retardation so that the induction of the syndrome in Down's syndrome patients adds a double reason for them to be retarded (*i.e.* patient B-IS-19 in the double blind study). Our patients developed symptoms in the appropriate age group and the pattern of seizures was very similar to other infantile spasm patients. The age of onset is 90% in the first year of life in classic infantile spasms (Jeavons, 1964) and 80% in the first year of life in the patients with '5-HTP induced' seizures that were classified as similar to classical infantile spasms. The social responsiveness of the child between seizures was affected, as also described in the infantile spasm patients (Jeavons and Bower, 1964; Illingworth, 1955). The association of the spasm with a cry, as noted in patients B-IS-19 and D-3 has also been reported in infantile spasms (Druckman and Chao, 1955; Jeavons and Bower, 1964).

The spontaneous frequency of seizures in untreated patients with Down's syndrome is difficult to ascertain, but must be considerably less than the 17% seen in our series of 5-HTP treated patients. Most series of infantile spasms patients included 1–2% Down's syndrome so there may be some predisposition to infantile spasms in this patient group. From the obverse point of view, however, the major texts on Down's syndrome (Benda, 1969; Penrose and Smith, 1966) do not even mention seizures. A large series from Russia reported some type of seizure phenomenon in 4.5% of the out-patient and 8.6% of an institutionalized population of mongols (Fedotov *et al.*, 1968). A recent neuropathological review of 26 cases of Down's syndrome disclosed seizures in childhood (age 3) in 1 patient and in the last year of life in 8 more patients in this series (Olsen and Shaw, 1969). Paulson reported that 12 out of 88 patients he studied with trisomy 21 had at least 1 seizure in later life. He commented that in no patient was there a history of infantile spasms (Paulson *et al.*, 1969). Seppalainen and Kivalo (1967) and MacGillivray (1967) found 8.7 and 8.1% of institutionalized patients had clinical seizures at some time in their lives. Seppalainen noted that epilepsy was most frequent in patients with congenital cardiac disease. Schachter (1956) and Richards *et al.* (1965) also found 8.8 and 4.4% of seizures in mongols. In Gibbs' extensive experience with EEGs, he recalled seeing 2 hypsarrhythmic EEGs in Down's syndrome patients (Gibbs *et al.*, 1964). We have had 1 patient with spontaneous infantile spasms in our Down's syndrome clinic; his 5-HI was no different than the trisomy mean. This rarity of seizures in young Down's syndrome patients, plus the high frequency (17%) of seizure phenomena in our 5-HTP treated series and their clinical cessation or amelioration of symptoms by decreasing 5-HTP in 7 of the 11 patients is strong suggestive evidence that chronic administration of 5-HTP was an etiological factor in the seizure activity reported in many of our patients.

According to Ellingson's recent reviews (Ellingson *et al.*, 1970; Ellingson, 1973) about 20–30% of randomly selected mongoloids display EEG abnormalities with most of the abnormalities occurring during childhood. Hypsarrhythmia, however, is a rare abnormality which is not mentioned in most reviews. It was found in only 2 of Gibbs *et al.* (1964) 84 mongoloid subjects. The incidence of seizures varies in different reports, *e.g.* in the Ellingson *et al.* (1970) series of 94 mongoloids, 23% had abnormal EEGs but none had seizures; in the Levinson *et al.* (1955) series of 40 mongoloid children 50% showed abnormal or borderline abnormal EEGs. Only 1 patient showed a hypsarrhythmic EEG and this child displayed seizures. Other series reported higher incidence of EEG abnormalities and seizures. Gibbs *et al.* (1964) reported that of the 184 mongoloids seen, 28% had abnormal EEGs. There were 18 patients (10%) with seizures. This high percentage was probably a reflection of the fact that this series came from a large EEG laboratory. Other sample biases, *e.g.* institutionalization as in Seppalainen and Kivalo's series (1967), may also account for some of the differences in reported incidence of EEG abnormalities and seizures. It is also apparent that mongoloids with seizures may have normal EEGs. A finding of a normal EEG in mongoloids who have seizures seems to be more frequent than in other retardates with seizures (Walter *et al.*, 1955).

The biochemistry of seizure phenomena is not fully understood. In most animal studies, low levels of 5-HT predispose to convulsions (Nellhaus, 1968) while anticonvulsants such as diphenylhydantoin raise 5-HT (Bonnycastle *et al.*, 1957). Furthermore, 5-HTP in animals raises seizure thresholds (Kobinger, 1958; Cooper *et al.*, 1968), In contrast, Coursin (1971) has reported a patient who had seizures induced by tryptophan.

There have been some studies of the biochemistry of patients with spontaneous infantile spasms. Of the abnormalities described in the literature, many can be linked to the 5-HT metabolic pathway—either to tryptophan, pyridoxine or 5-HT (Schlesinger *et al.*, 1968; Leins *et al.*, 1959; Bower, 1961;

Low *et al.*, 1958). We now add 5-HTP so that each step of the pathway leading to 5-HT has been indited (Fig. 1-1).

Examination of 5-HT levels in children with 'spontaneous' infantile spasms of the many different etiologies other than '5-HTP induction', has shown that 46% (Ota, 1969) to 50% (Coleman, 1971) have elevated 5-HI levels in the blood. However, these high platelet levels were accompanied by a very low CNS level in the 1 patient who has come to autopsy. Certainly most of the animal literature suggests that low CNS levels accompany seizure phenomena so this result is not unanticipated.

To try to fit our results in with these concepts one could postulate that our patients who developed seizures on 5-HTP actually had depressed CNS levels of serotonin due to 5-HTP interference with the active transport of tryptophan, the natural precursor of 5-HT, into the brain. However, with the multiple CNS biochemical abnormalities seen in patients with an entire extra chromosome, any conclusion is very qualified.

It is doubtful that 5-HT levels, either high or low, are solely responsible for seizures in these patients or any patients. It is likely that clinical phenomena, such as a seizure, are related to an interruption of the precise balance between several biochemical modalities, rather than an abnormal level of a single endogenous amine.

Infantile spasms are age related seizure syndromes not seen in older children and adults. Although the induction of these phenomena raises serious questions about the use of 5-HTP in infants with Down's syndrome, its safety in other infants not already prone to infantile spasms, and in older children and adults remains to be clarified. As of the date of publication of this monograph, we have given 5-HTP to 6 children aged 8 years to 13 years, with severe behavior disorders of childhood, for between 6 months and 16 months without any seizure phenomena clinically or in EEG tracings. In this patient group, who were on a dosage of 1.1 mg/kg to 10.7 mg/kg, only acute and reversible side effects (insomnia, irritability, hyperreflexia) have been recorded.

The ophthalmologists treating the two patients with glaucoma felt the evidence was primarily in favor of congenital, non-drug related origin to the symptoms in each case. However, we were troubled by the fact that 1 patient had an identical twin who did not receive 5-HTP and did not develop glaucoma, and by the fact that we were unable to establish the spontaneous incidence of infantile glaucoma in Down's syndrome from the literature; apparently it is not one of the signs usually seen. It is not mentioned in ocular surveys of Down's syndrome (Ormond, 1912; Lowe, 1949; Eissler and Lougenecker, 1962). There is only one unequivocal, non-familial case in the literature—in the Hall study of 38 trisomy patients born in a specific geographical area of Sweden; this study listed 1 patient with congenital glaucoma (Hall, 1964). The Gardiner (1967) review of 22 Down's syndrome children in an institution reported that 1 patient had choroido-retinal atrophy, a finding associated with several conditions, including congenital glaucoma. There is also a patient with trisomy 21, bilateral congenital glaucoma and keratoconus reported in the literature who is a member of a family that contains 11 members suffering from Rieger's anomaly, many of whom had associated glaucoma (Dark and Kirkham, 1968). Keratoconus, which may be associated in the acute phase with 'secondary glaucoma', was found in patients in the older age groups in one institution (Cullen and Butler, 1963).

The presence of only 2 cases in the entire literature (1 familial) compared to 2 cases in our series of 62 5-HTP treated cases raises the possibility that 5-HTP administration may be a factor in glaucoma in our series.

Increased rate of pupillary dilation to atropine (Berg *et al.*, 1959; O'Brien *et al.*, 1960; Mir and Cumming, 1971) has been reported in Down's syndrome patients; this ocular variation could be theoretically related to 5-HT metabolism. It has no apparent clinical significance.

Benda (1960) reported "partial or complete baldness is not rare" in Down's syndrome. Therefore it is impossible to draw any conclusions regarding 5-HTP side effects from the alopecia areata of 1 patient (D-8) even though her hair began to regrow when she was taken off 5-HTP. 5-HTP's cousin amino acid—L-Dopa—may be a factor in control of hair growth; loss was seen in older women (Marshall and Williams, 1971) and regrowth in older men. Endocrine factors affect hair growth and 5-HTP and L-Dopa are the precursors of amines which indirectly regulate hypothalamic function. An endocrine study of patient D-8 during her period of alopecia areata was within normal limits.

In our clinic, 5-HTP has never been tried on a pubertal Down's syndrome female so we have no information on its effects on the menses of this patient group. An effect on menstruation by L-Dopa occurred in both Parkinson (Ansel, 1970) and dystonia (Coleman, unpublished data; Chase, unpublished data) patients; this effect was predicted by Sandler's studies on the L-Dopa increase of monoamine oxidase in the rat uterus (Sandler, 1970). No similar studies have been done with 5-HTP. The only pubertal female to whom we have given 5-HTP is a 14-year-old with a diagnosis of the idiot savant syndrome, low endogenous 5-HI levels and markedly increased 5-HT efflux from the platelets (Appendix VII-6). She had a doubling of the length of her first 2 menstrual periods after starting L-5-HTP at 8 mg/kg; by the third month of administration of the amino acid, an accommodation occurred and her periods returned to their pre-treatment length.

The biochemical effects of long-term administration of 5-HTP to human beings is unknown. Particularly when starting in the neonatal period, it must be assumed that enzymes, transport and binding mechanisms may be altered. Weiss *et al.* (1971) have shown that massive dosages of L-Dopa interfere with protein synthesis by disaggregation of polysomes, unbinding RNA from ribosomes. Comparable studies were not done on 5-HTP. If similar effects are seen with 5-HTP and its relevance to humans established, this could be a contraindication to the amino acid in young children. Another L-Dopa study has shown that large doses of L-Dopa will decrease the L-aromatic amino acid decarboxylase level in liver up to 50% without changing the level in brain or kidney. The presence or absence of B6 was not a factor. This is the same enzyme utilized by 5-HTP to make 5-HT and studies of large dosages of 5-HTP would be of interest.

In general, experimental trials of large doses of 5-HTP should be used with great caution in young children, particularly if the disease entity under investigation is one with a known incidence of the infantile spasm syndrome.

REFERENCES

ANSEL, R. D. (1970) General discussion in L-Dopa and parkinsonism, Eds. A. Barbeau and F. H. McDowell, Philadelphia, p. 317. F. A. Davis Comp., Philadelphia, Pa.

BAZELON, M., BARNET, A., LODGE, A. and SHELBURNE, S. (1968) The effect of high doses of 5-hydroxytryptophan on a patient with trisomy 21, clinical chemical and EEG correlations. *Brain Research*, **11**, 397.

BENDA, C. E. (1960) The child with mongolism (congenital acromicria), Grune and Stratton, New York.

BENDA, C. E. (1969) Down's syndrome—mongolism and its management, Grune and Stratton, New York.

BERG, J. M., BRANDON, M. and KIRMAN, B. H. (1959) Atropine in mongolism. *Lancet*, **2**, 441.

BONNYCASTLE, D. D., GIARMAN, N. J. and PAASONEN, M. K. (1957) Anticonvulsant compounds and 5-hydroxytryptamine in rat brain. *Brit. J. Pharmacol.* **12**, 228.

BOWER, B. D. (1961) The tryptophan load test in the syndrome of infantile spasms with oligophrenia. *Proc. Roy. Soc. Med.* **54**, 540.

COLEMAN, M. (1970) Preliminary remarks on the L-Dopa therapy of dystonia. *Neurology*, **20**, 114.

COLEMAN, M. (1971) Infantile spasms induced by 5-hydroxytryptophan in patients with Down's syndrome. *Neurology*, **21**, 911.

COOPER, A. J., MOIR, A. T. B. and GULDBERG, H. C. (1968) The effect of electroconvulsive shock on the cerebral metabolism of dopamine and 5-hydroxytryptamine. *J. Pharm. Pharmacol.* **20**, 729.

COTZIAS, G. C. (1968) L-Dopa for Parkinsonism. *New Engl. J. Med.* **278**, 630.

COURSIN, D. B. (1971) Central nervous system hypersensitivity to tryptophan. *Am. J. Clin. Nutr.* **24**, 821.

CULLEN, J. F. and BUTLER, H. G. (1963) Mongolism (Down's syndrome) and keratoconus. *Brit. J. Ophthal.* **47**, 321.

DARK, A. J. and KIRKHAM, T. H. (1968) Congenital corneal opacities in a patient with Rieger's anomaly and Down's syndrome. *Brit. J. Ophthal.* **52**, 631.

DRUCKMAN, R. D. and CHAO, D. H. (1955) Massive spasms in infancy and childhood. *Epilepsia*, **4**, 61.

EISSLER, R. and LOUGENECKER, L. P. (1962) The common eye findings in mongolism. *Am. J. Ophthal.* **54**, 398.

ELLINGSON, R. J. (1973) EEG disorders associated with chromosome anomalies. In: International handbook of EEG and clinical neurophysiology, **13**, Elsevier, Amsterdam, in press.

ELLINGSON, R. J., MENOLASCINO, F. J. and EISEN, J. D. (1970) Clinical-EEG relationships in mongoloids confirmed by karyotype. *Am. J. Ment. Defic.* **74**, 645.

ERSPAMER, V. (1966) 5-hydroxytryptamine and related indolealkylamines, Springer-Verlag, New York.

FEDOTOV, D. D., SHAPIRO, Y. L. and VAINDRUKH, F. A. (1968) Some features of paroxysmal states in Down's disease (clinico-statistical analysis). *Zh. Nevropat. Psikhiat. Korsakov.* **68**, 1516.

GARATTINI, S. and VALZELLI, L. (1965) Serotonin, Elsevier, Amsterdam.

GARDINER. P. A. (1967) Visual defects in cases of Down's syndrome and in other mentally handicapped children. *Brit. J. Ophthal.* **51**, 469.

GIBBS, E. L., GIBBS, F. A. and HIRSCH, W. (1964) Rarity of 14- and 6-per-second positive spiking among mongoloids. *Neurology*, **14**, 581.

HALL, B. (1964) Mongolism in newborns. *Acta Pediatrica* (supplement), **154**, 3.

HAVERBACK, B. J. and DAVIDSON, J. D. (1958) Serotonin and the gastrointestinal tract. *Gastroenterology*, **35**, 570.

ILLINGWORTH, R. S. (1955) Sudden mental deterioration with convulsions in infancy. *Arch. Dis. Child.* **39**, 529.

JEAVONS, P. M. and BOWER, B. D. (1964) Infantile spasms, William Heinemann Ltd, London.

KOBINGER, W. (1958) Influence of changes of serotonin content of the central nervous system on the seizure threshold of cardiazol. *Naunyn Schmiedeberg Arch. Pharm.* **233**, 559.

LEINS, M., COTTE, J., HERMIER, M., YASSE, L. and LENICHE (1959) L'épreuve de charge au tryptophane comme moyen de détection des apyridoxinoses chez l'enfant. *Pédiatrie*, **14**, 853.

LEVINSON, A., FRIEDMAN, A. and STAMPS, F. (1955) Variability of mongolism. *Pediatrics*, **16**, 43.

LOW, N. L., BOSMA, J. F., ARMSTRONG, M. D. and MADSEN, J. D. (1958) Infantile spasms with mental retardation: I. Clinical observations and dietary experiments. *Pediatrics*, **22**, 1153.

LOWE, R. F. (1949) The eyes in mongolism. *Brit. J. Ophthal.* **33**, 131.

MACGILLIVRAY, R. C. (1967) Epilepsy in Down's anomaly. *J. Ment. Defic. Res.* **11**, 43.

MARSHALL, A. and WILLIAMS, M. J. (1971) L-DOPA and alopecia. *Brit. Med. J.* **2**, 47.

MIR, G. H. and CUMMING, G. R. (1971) Response to atropine in Down's syndrome. *Arch. Dis. Child.* **46**, 61.

MULLER, S. A. and WINKELMANN, R. K. (1963) Alopecia areata: an evaluation of 736 patients. *Arch. Derm.* **88**, 106.

NELLHAUS, G. (1968) Relationship of brain serotonin to convulsions. *Neurology*, **18**, 298.

O'BRIEN, D., HAAKE, M. W. and BRAID, B. (1960) Atropine sensitivity and serotonin in mongolism. *Am. J. Dis. Child.* **100**, 873.

OLSEN, M. I. and SHAW, C. M. (1969) Presenile dementia and Alzheimer's disease in mongolism. *Brain*, **92**, 147.

ORMOND, A. W. (1911–1912) Notes on the ophthalmic condition of forty-two mongolian imbeciles. *Trans. Opthal. Soc. of the U.K.* **32**, 69.

OTA, S. (1969) Study of serotonin metabolism in pediatrics. 2. Blood serotonin levels in various diseases in children. *Acta Paediatrica Japonica*, **73**, 61.

PARTINGTON, M. W., MACDONALD, M. R. A. and TU, J. B. (1971) 5-Hydroxytryptophan (5-HTP) in Down's syndrome. *Develop. Med. Child Neurol.* **13**, 362.

PAULSON, G. W., SON, C. D. and NANCE, W. E. (1969) Neurological aspects of typical and atypical Down's syndrome. *Dis. Nerv. Syst.* **30**, 632.

PENROSE, L. S. and SMITH, G. F. (1966) Down's anomaly, J. and A. Churchill Ltd, London.

POLLIN, W., CARDON, P. V. and KETY, S. S. (1961) Effects of amino acid feedings in schizophrenic patients treated with iproniazid. *Science*, **133**, 104.

RICHARDS, B. W., STEWART, A., SYLVESTER, P. E. and JASIEWICZ, V. (1965) Cytogenetic survey of 225 patients diagnosed clinically as mongols. *J. Ment. Defic. Res.* **9**, 245.

Sandler, M. (1970) General discussion in L-Dopa and Parkinsonism, Eds. A. Barbeau and F. H. McDowell, Philadelphia, p. 317.

Saxl, O. (1968) Down's syndrome: new aspects. *Acta Paediat. Acad. Scient. Hung.* **9**, 329.

Schacher, M. (1956) Les convulsions chez les mongoliens. *Med. Infant.* **63**, 5.

Schlesinger, K., Boggan, W. and Freedman, D. X. (1968) Genetics of audiogenic seizures. II. Effects of pharmacological manipulation of brain serotonin, norepinephrine, and gamma-aminobutyric acid. *Life Sci.* **7**, 437.

Seppalainen, A. M. and Kivalo, E. (1967) EEG findings and epilepsy in Down's syndrome. *J. Ment. Def. Res.* **11**, 116.

Walter, R. D., Yeager, C. L. and Rubin, H. K. (1955) Mongolism and convulsive seizures. *Arch. Neurol. Psychiat.* **74**, 559.

Weiss, B. F., Munro, H. N. and Wurtman, R. J. (1971) L-DOPA; disaggregation of brain polysomes and elevation of brain tryptophan. *Science*, **173**, 833.

CHAPTER 8

Other Methods of Raising Serotonin

MARY COLEMAN

Serotonin can be raised in the central nervous system by a great variety of drugs; some of them are listed in Appendix IX-1. When we were considering a long-term attempt to elevate 5-HT in the low serotonin syndromes, four possibilities other than 5-HTP were explored; monoamine oxidase inhibitors, vitamin B6, tryptophan and imipramine. Each of these possibilities is a non-specific way of raising serotonin because amines other than 5-HT also are affected.

Early reports of the effect of administering other 5-HT elevating drugs in patients with Down's syndrome are described in this chapter.

Monoamine Oxidase Inhibitors

It has long been established that monoamine oxidase inhibitors (MAOI) result in the accumulation of 5-HT in the central nervous system (Spector *et al.*, 1958). They also increase 5-HT levels in platelets in most groups of patients studied in the literature. Many studies with MAOI have been with adult patient groups. However, Kirman and Pare (1961) treated 4 children with PKU, one of the low 5-HT syndromes, with 50 mg of iproniazid daily for 6 months. A rise in serum levels of 5-HIAA was noted. The patients were between 7 and 15 years and had no change in mental functioning during the 6-month treatment period.

Five investigators have given the monoamine oxidase inhibitor, nialamide, to patients with Down's syndrome. It has been reported that a 35-year-old mongol who was mute began to speak and showed increased physical activity. Rett (1959) reported improvement in motor activity in young children who began to either stand or walk. Vasquez and Turner (1951) also reported "quickened mental and physical" reactions although they noted there were no I.Q. changes. Heaton-Ward (1962) then did a controlled study with 51 institutionalized patients and found no sustained improvement as tested by a psychologist, although the nursing staff saw some improvement in patients on the drug. Perry (1962) gave 100 mg/day to 5 patients with Down's syndrome (ages unspecified) and reported increases of serotonin, tryptamine, normetanephrine and *p*-tyramine in their urines.

Benson and Southgate (1971) demonstrated that platelet monoamine oxidase activity is significantly lower in patients with Down's syndrome, although Lott *et al.* (1972) do not confirm this. If MAO is depressed, any inhibitor has less substrate to affect, at least in the platelets. This low level of MAO has been demonstrated in the cell (platelet) that contains the 5-HI we measured by our whole blood method. Benson and Southgate possibly thought they could prove that the low 5-HI levels in blood were due to increased MAO, since

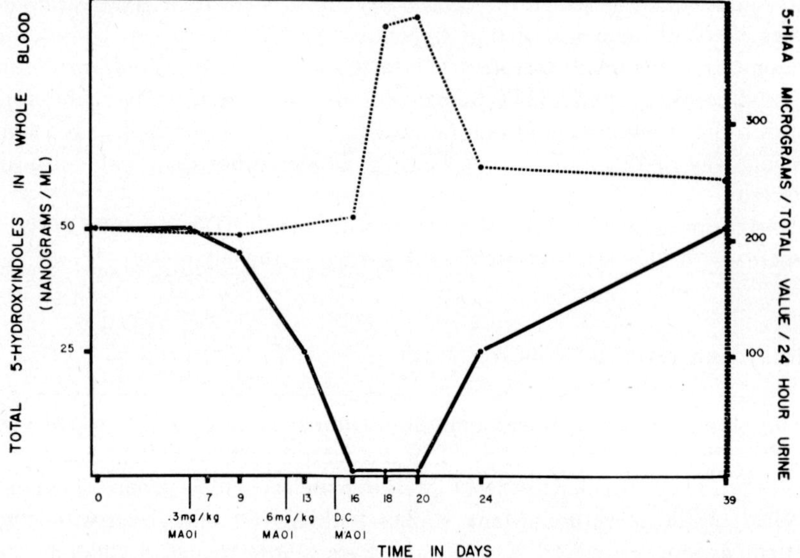

Fig. 8-1. Tranylcypromine administration in an 18-day-old patient with trisomy 21 resulted in opposite effects on whole blood 5-HI and urinary 5-HIAA levels.

many other intracellular enzymes are elevated in Down's syndrome children. Instead they found a low level of the enzyme. This helps explain the results of Paasonen *et al.* (1964) who gave iproniazid and isocarboxozid to Down's syndrome patients from 5 to 13 years of age and found that they elevated platelet 5-HT for a much shorter period than in other patients.

We elected to give a non-hydrazine monoamine oxidase inhibitor, tranylcypromine, to a neonatal patient with the trisomy 21 form of Down's syndrome. We, of course, assumed tranylcypromine would raise the patient's whole blood level of 5-HI and urinary 5-HIAA. We began tranylcypromine when the patient was 18 days of age. To our distress (but not, apparently, the patient's) blood 5-HI dropped to zero and remained there for 7 days following the last dose of the MAOI (see Fig. 8-1). Urinary 5-HIAA rose to its maximum during the period when no blood level of 5-HI could be detected. The clinical examination was unchanged during this period, in particular no increase in hypotonia or change in blood pressure was recorded. The patient was hypotonic before and he remained so.

Did this paradoxical fall of blood 5-HI occur because the patient was so young or because the patient had Down's syndrome? We could find no literature on MAOI given to neonates.

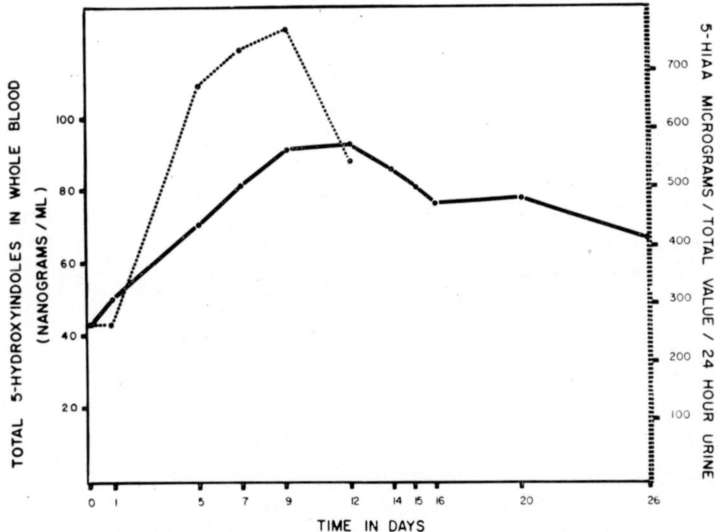

Fig. 8-2. Tranylcypromine administration in a 9-year-old patient with trisomy 21 resulted in a rise of both whole blood 5-HI and urinary 5-HIAA levels.

We felt that it would not be possible to repeat this experiment on another Down's syndrome infant whose intellectual potential was not yet established. However, we elected to give tranylcypromine to 2 older children of 3 and 9 years of age with trisomy karyotype, in a study to check the literature reporting rising 5-HT in Down's syndrome children on MAOI (Perry, 1962; Paasonen *et al.*, 1964).

In both the older children, whole blood 5-HI rose with the administration of tranylcypromine in mg/kg dosages comparable to the infant patient (see Fig. 8-2). The trial in the 3-year-old patient had to be terminated when the child's 5-HI level reached $2\frac{1}{2}$ times the baseline level. At that point he became too 'turned on', babbling constantly during the day and laughing, playing, euphoric with profound insomnia at night. He also had reflexes brisker than baseline during this period. The 9-year-old patient had a higher baseline 5-HI level at the beginning of the experiment and his highest value on tranylcypromine was double his baseline. There was very little change in his behavior or neurological examination.

Why did the paradoxical fall of 5-HI mirrored conversely by a rise in 5-HIAA occur in the infant patient? It appears on the basis of this limited study to be an age-related, rather than mongol-related phenomenon. Did tryptamine, a less pharmacologically active amine elevated by MAOI (Sjoerdsma *et al.*, 1959) displace binding sites in the platelets giving a 'false neurotransmitter' effect? (Kopin *et al.*, 1965; Roberge *et al.*, 1971). The other established pathways of 5-HT metabolism that are known to be activated when MAO inhibition occurs (serotonin-*O*-glucuronide, *N*-acetylserotonin) do not go to 5-HIAA. Could 5-HT have been metabolized to 5-HIAA through 5-hydroxyindolepyruvic acid (not demonstrated in humans) perhaps by the enzyme with weak amine oxidation ability described in brain tissues by Weissbach *et al.* (1961)?

Because of the unexpected 5-HT drop to zero in our 18-day-old patient, monoamine oxidase inhibitors should be used with great caution, if at all, in infant patients with Down's syndrome.

Vitamin B6

Vitamin B6, a substituted pyridine ring, occurs in natural foods as pyridoxal, pyridoxine and pyridoxamine. In the body, these precursors are phosphorylated by an ATP dependent kinase to pyridoxal-5-phosphate (PLP) or pyridoxamine-5-phosphate, the active co-enzyme forms. In this monograph, the term 'vitamin B6' is used in a generic sense. It refers to the combined biologic effect of all chemical forms of the vitamin which may be present—those occurring in natural foods (pyridoxine, pyridoxal, pyridoxamine) and phosphates of the latter two as formed in the body.

Some investigators have produced evidence suggesting that there may be a decreased amount of vitamin B6 in Down's syndrome. The demonstration of a deficiency in a group of retarded patients is of concern because there is evidence in newborn animals that pyridoxine deficiency selectively will decrease brain enzyme functions earlier than similar enzyme systems outside the central nervous system (Valadares, 1967). In short, during the early weeks of life (the prime age of risk in mental retardation) the brain may be the organ most affected by a vitamin B6 deficiency.

Vitamin B6 is one of the most ubiquitous co-enzymes in the body; it is involved in the metabolism of proteins, lipids, carbohydrates, purines and pyrimidines and is also needed in the synthesis of five other co-enzymes. In the brain, vitamin B6 helps the active transport of many amino acids into the central nervous system as well as functioning as a co-enzyme in these metabolic pathways including all compounds implicated to date as possible neurotransmitters.

The vitamin is the co-enzyme of the second step of the 5-HT pathway (see Fig. 1-1). Its apo-enzyme in this pathway, L-aromatic amino acid decarboxylase (L-AAAD), is the pyridoxine dependent enzyme that in addition to transforming 5-HTP to 5-HT synthesizes a number of central amines, including dopamine of the catecholamine pathway.

However, the serotonin pathway appears to have a slight advantage over the other central amine pathways because 5-HTP is the substrate with the lowest Km (that is, greatest affinity) for L-AAAD (Lovenberg, 1970). Also, although it is known that the decarboxylase enzymes are more likely to be affected by vitamin B6 deficiency than other B6 dependent enzymes such as the transaminases (Snell, 1958), L-AAAD appears to be more tightly bound to the co-enzyme than many of the other decarboxylases.

Could vitamin B6 deficiency be another factor causing the low 5-HI levels we consistently measure in the Down's syndrome patients? It is known that the vitamin is involved in at least two steps of 5-HT metabolism: (1) increasing the transport of tryptophan, the usual precursor of 5-HT, into cells, and (2) being the co-enzyme of the L-AAAD step in the 5-HT pathway.

In the last 10 years, a number of papers have been published related to vitamin B6 metabolism in patients with the trisomy 21 form of Down's syndrome. Some of the results have been interpreted to suggest a 'deficiency' syndrome in patients with the trisomy 21 form of Down's syndrome. Abnormal tryptophan loading tests have been cited as evidence of diminution of vitamin B6 levels in patients with Down's syndrome (Gershoff *et al.*, 1958; O'Brien and Groshek, 1962; Jérôme, 1962). However, there is disagreement about the value of tryptophan loading tests as predictors of vitamin B6 deficiency (Hansson, 1969). In a separate study of tryptophan loading tests, McCoy and Chung (1964) did not find their results initially significant, but were able to demonstrate a difference when Down's syndrome and control groups were pretreated with a vitamin B6 antagonist, deoxypyridoxine. Then, in a further study of the excretion of the end-product of vitamin B6 metabolism, 4-pyridoxic acid (4-PA), they concluded that Down's syndrome children have "a smaller dissociable pool of vitamin B6", probably due to "the enzymes of vitamin B6 catabolism having greater activity in mongoloid subjects than controls" (McCoy *et al.*, 1968). A recent study of tyrosine and tryptophan pathways in Down's syndrome also suggests deficient functioning of pyridoxine dependent enzymes in institutionalized patients (Kontsevaya and Ritsner, 1972).

Direct measurements of pyridoxal-5-phosphate (PLP) in the leukocytes of

20 institutionalized patients with Down's syndrome compared to other retarded patients in the same institution, as well as normal controls, were done by Coburn and Seidenberg (1968). They found a statistically significant depression of PLP levels in trisomy 21 patients compared to retarded patients from the same institution. However, in our study of PLP levels in home-reared trisomy 21 patients, comparing their levels to controls and siblings, Bhagavan *et al.* (1973) were unable to confirm unequivocal evidence of PLP deficiency.

There also is a possibility of pyridoxine 'dependency' in Down's syndrome. Low levels of taurine, formed by a pyridoxine dependent pathway, have been reported in the urine of this patient group (Goodman *et al.*, 1964). Yet, administration of moderate doses (2 mg/kg) of vitamin B6 does not alter taurine levels (King *et al.*, 1968).

A number of physicians have included vitamin B6 in their multiple vitamin–mineral preparations they recommend in Down's syndrome patients. None of these 'therapies' have had an independent evaluation that supported the original claims to date.

1. Goldstein (1954) used B6 (dose unclear) as a 'muscle relaxant' and pointed out that it was utilized in pathways involving protein metabolism; he also felt that it helped the seborrheic dermatitis of the face and scalp. There is no independent study by another author of these recommendations.

2. Haubold *et al.* (1960) included a 27 mg/100 ml daily dose of vitamin B6 in his extensive vitamin–mineral–hormonal regimen for Down's syndrome patients which also included 'cellular' therapy. White and Kaplitz (1964) compared the Haubold treatment (minus the cellular therapy portion) to placebo and were unable to demonstrate any effect neurologically or on intelligence quotients. In particular, they noted little change in hypotonicity. The preparation may have resulted in increased aimless, gross motor activity.

3. Turkel (1963) includes 20 mg of vitamin B6 (or less in younger patients) in his daily vitamin–mineral preparation for Down's syndrome children. His report of improvement in his patients has been disputed by Bumbalo *et al.* (1964) who tried the Turkel preparation on another group of patients.

4. Tanino (1969) of Japan included vitamin B6 at dosages of 30–70 mg/day in his multiple vitamin cocktail for Down's syndrome. There is no independent evaluation of Tanino's therapy available.

There is an interesting patient in the literature with a translocation karyotype who was given vitamin B6 (Carter, 1967). [Translocation patients occasionally can function as normal individuals; there are 2 such patients in the literature with an I.Q. of 80 (Shaw, 1962) and 82 (Finley *et al.*, 1965).] The patient was given 300 mg of vitamin B6 from the third week of life and had unusually dedicated parents. At 4 years of age, the patient had an estimated I.Q. of 120 (Cattel Infant Scale); now she is in a normal classroom. The physician, who also tried B6 unsuccessfully in other patients, does not feel that the vitamin necessarily had any effect in this case (C. H. Carter, personal communication).

We have openly administered vitamin B6 to 71 patients, starting in the neonatal period in two-thirds of the cases, with Down's syndrome; 23 in combination with 5-HTP (discussed in Chapter 3) and 48 without 5-HTP. Because of falling 5-HI levels, tryptophan was added later to the regimen of 18 in this latter group. In addition to open studies, a double blind study of large doses of vitamin B6 started in the neonatal period has been initiated.

In any patient placed on the vitamin shortly after visiting our clinic, a rise in 5-HI could be attributed to the placebo effect. One approach to compensate for the placebo effect was to select patients for study who were already familiar with the clinic. Patient B-2, a trisomy 21 child on the 5-HTP/placebo double blind study, was chosen for a trial of vitamin B6 starting at 3 years of age, just after the double blind study was terminated. In the previous 2 years, while he was on placebo, his 5-HI levels averaged 54 ng/ml (age 1–2) and 56 ng/ml (age 2–3).

He was placed on 5 mg/kg of vitamin B6 at 37 months of age and the dosage gradually increased through the year to 12 mg/kg (Appendix VIII-1). The 5-HI level varied as much as 50 ng/ml in the first months but then settled to a continuous level significantly higher than the 2 baseline years (see Fig. 8-3).

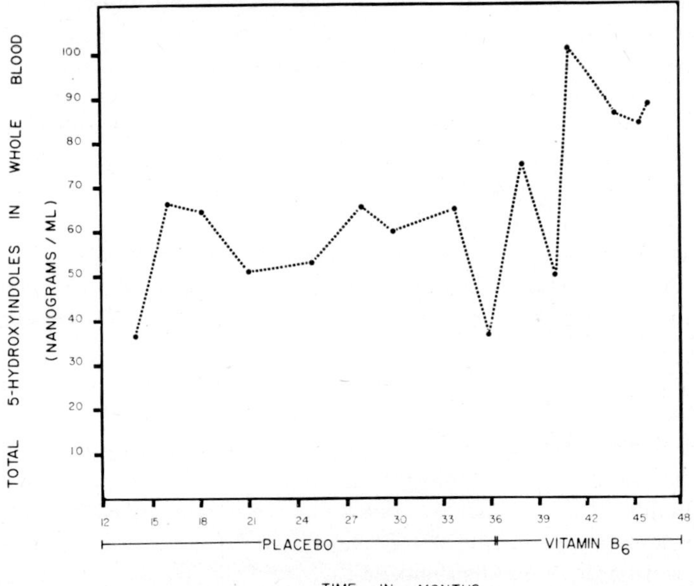

Fig. 8-3. Trisomy 21 patient B-2 was on placebo between 12–36 months of age and received vitamin B6 between 36–48 months of age. Except for 1 overlapping value, the levels of whole blood 5-HI were higher during the period of vitamin B6 administration.

This pattern of a drop in 5-HI level prior to the rise to a higher baseline has been seen in a number of older patients placed on B6. During his 3 baseline years, patient B-2's parents had known he had a 50% chance of being on a research medicine; during the B6 year, they knew he was definitely on a research drug. Possibly this was a factor in the elevation of 5-HI during the year B6 was administered. Patient C-33, a patient with a double chromosomal error and a high first visit 5-HI (148 ng/ml), was placed on vitamin B6 and had an elevation of 5-HI above the normal range. Since that experience, we give vitamin B6 only to Down's syndrome patients with low 5-HI levels. Conversely, we also have seen 5-HI values decrease when several patients were taken off B6.

Another approach to evaluate the effects of B6 on 5-HI levels is to study new clinic patients and compare them with matched controls. One such survey was completed consisting of first visit levels in 17 patients who were seen in our clinic *after* being placed on B6 by other physicians. All these patients were trisomy 21 and under 1 month of age when first seen in our clinic. This would be a study of acute effects in the neonatal period because B6 had only been administered a short period of time. A comparison of their 5-HI values with available age-matched controls selected from our basic first visit population (Appendix I-2) disclosed that the mean 5-HI value was 40 ng/ml for the controls and 62 ng/ml for patients receiving B6, a statistically significant increase (Appendix VIII-2). However, because in most cases the 5-HI values of the 2 groups were determined on different laboratory determinations, and because the parents of B6 patients knew they were receiving a research medicine, this is not conclusive evidence of a vitamin B6 effect.

However, the literature suggests that it may be correct. In normal newborns, Berman *et al.* (1965) has shown elevation of whole blood 5-HT levels (determined with a micro spectrofluorometric procedure) by giving vitamin B6 either once a day or in each bottle. He and his collaborators also studied low birth weight infants, a group more comparable to Down's syndrome patients. Although less marked, a rise in the whole blood 5-HT levels was also seen in this group.

Tu̦ and Zellweger (1965) gave vitamin B6 and a B6 antagonist (D,L-penicillamine) in older Down's syndrome patients with low whole blood 5-HT levels and tested their reaction to tryptophan loading. Their studies suggested that there was a relative depression of the metabolic step (decarboxylation) using vitamin B6 as a co-factor. After giving the B6 antagonist, they were able to show a statistically significant rise in blood 5-HT levels by the combined administration of tryptophan and B6.

These studies are in accord with animal experiments showing pyridoxine deficient diets have resulted in depressed 5-HT levels (Weissbach *et al.*, 1957; Beiler, 1954).

Coburn (personal communication) has given up to 200 mg pyridoxine/day to adolescents with Down's syndrome in a preliminary study. He failed to demonstrate any effect on blood 5-HT levels but is continuing further studies.

McKean (in preparation) has done a hallmark turnover study of the effect of vitamin B6 on serotonin synthesis in the central nervous system of patients with Down's syndrome. He states: "The administration of 100 mg of pyridoxine HCl per day in 3 infants from 2 weeks to 4 months of age induced a significant increase in the rate of serotonin synthesis as measured by probenicid induced accumulation of 5-HIAA in the cerebral spinal fluid of such patients."

The toxic effects of vitamin B6 appeared, in part, to mimic those of 5-HTP in a number of ways. The most serious side effect of 5-HTP (the infantile spasm syndrome) occurred in 2 patients receiving vitamin B6 although the percentage of patients affected (1.3% was considerably lower than in the 5-HTP-induced series. One patient, a translocation aged 7 months, developed the full-blown syndrome with hypsarrhythmic EEG. Stopping the vitamin B6 (a very low dose of 0.5 mg/kg) had no discernible effect and ACTH was needed to reverse the disease process. In the second patient, stopping the vitamin reversed the intermittent seizures and borderline EEG findings. It is possible that infantile spasms would have occurred spontaneously in these B6 patients, particularly in the first patient who was on an extremely low dose of the vitamin.

Down's syndrome patients have a higher than usual frequency of spontaneous infantile spasms. In our clinic we have only been referred 1 patient who had spontaneous infantile spasms. After the usual therapeutic measures failed, a trial of vitamin B6 was given briefly. It had to be stopped because the frequency of seizures doubled.

There is an extensive literature relating vitamin B6 to seizures. Most of it related B6 deficiency to seizures; B6 excess also has been reported (Wiechert, 1969; Gallagher, 1971). It is generally believed that the effect of vitamin B6 on seizure thresholds may not be related to the 5-HT pathway but to the glutamic pathway. Giving vitamin B6 results in differential stimulation of the two enzyme systems—GA decarboxylase and GABA aminotransferase—with a resulting change in the relative concentrations of GA and GABA (Bhagavan and Coursin, 1971).

Other side effects of B6 similar to 5-HTP are insomnia and hyperactivity. Insomnia is helped by giving most or all of the dose in the morning. Hyperactivity can be controlled by lowering the dosage of the vitamin. Recently, a study of large doses of pyridoxine given to young rats has shown increased body weight (Cohen *et al.*, 1973). If this increase also applies to length, it might not be a negative side effect if applied to Down's syndrome patients.

Vitamin B6 may raise 5-HT levels in other disease entities also. In one of our neurology patients with spontaneous infantile spasms and high 5-HI levels in whole blood (Coleman, 1971), a trial of B6 resulted in even higher blood 5-HI levels. Two other patients, with still undiagnosed forms of mental retardation associated with low 5-HI levels, have also received vitamin B6 with subsequent elevation of the whole blood 5-HI levels.

The studies, showing a rise in 5-HT following vitamin B6 administration, raise more questions than they answer. If the first step of the 5-HT pathway (Fig. 1-2) is rate-limiting (Macon *et al.*, 1971), why does the co-enzyme of the second step affect the amount of end product? Perhaps the first enzyme is rate-limiting in usual situations, but the second step can assume that role in some diseases or in the presence of abnormally large amounts of the vitamin B6 co-enzyme. There is some experimental data to show that the concentration of added pyridoxal phosphate in homogenates of guinea pig kidney can contribute to a feed-back mechanism regulating 5-HT (Contractor and Jeacock, 1967). A relationship between the activity of pyridoxal kinase and brain concentrations of 5-HT also has been shown (Ebadi *et al.*, 1968). Ebadi (1970) has recently stated, "The concentration of PLP (pyridoxal leukocyte phosphate) might control the production of biogenic amines in the brain."

Tryptophan

There is good evidence that one major factor controlling brain serotonin levels is plasma tryptophan levels (Fernstrom and Wurtman, 1971) and some recent data to show that one mechanism for a drug (Librium) to affect 5-HT is by increasing the levels of brain tryptophan (Perez-Cruet *et al.*, 1971).

In Down's syndrome patients, there is evidence of less efficient intestinal absorption of tryptophan (Jérôme *et al.*, 1960; Jérôme, 1962). O'Brien and Groshek (1962) have shown that blood tryptophan levels are lower in this patient group.

For these reasons, a limited study of the addition of tryptophan to the diet of 18 Down's syndrome children is just beginning in our clinic. The tryptophan is always given with vitamin B6 so it is not a study of the amino acid alone. Only if a definite effect is noted when the patients reach 3 years of age will separate, component studies be undertaken using blind controlled conditions. Blood 5-HI levels appear to be increasing, based on fragmentary data.

Tricyclic Antidepressants

A trial of imipramine HCl, a member of the dibenzazepine group of compounds and a tricyclic antidepressant, was initiated with a 9-year-old patient who has the trisomy 21 form of Down's syndrome. Several investigators had noted that this compound or its metabolite, desipramine, could cause an increase in 5-HT concentration in animal brains (Kivalo *et al.*, 1961; Segawa, 1970). One mechanism of the increase may be inhibition of the uptake by synaptic vesicles of 5-HT formed within neurons, that is, it may block the intracellular 5-HT concentrating mechanisms.

When placed on imipramine, the patient's whole blood 5-HI levels sank to less than one-half of baseline (see Fig. 8-4). It could be argued that there was

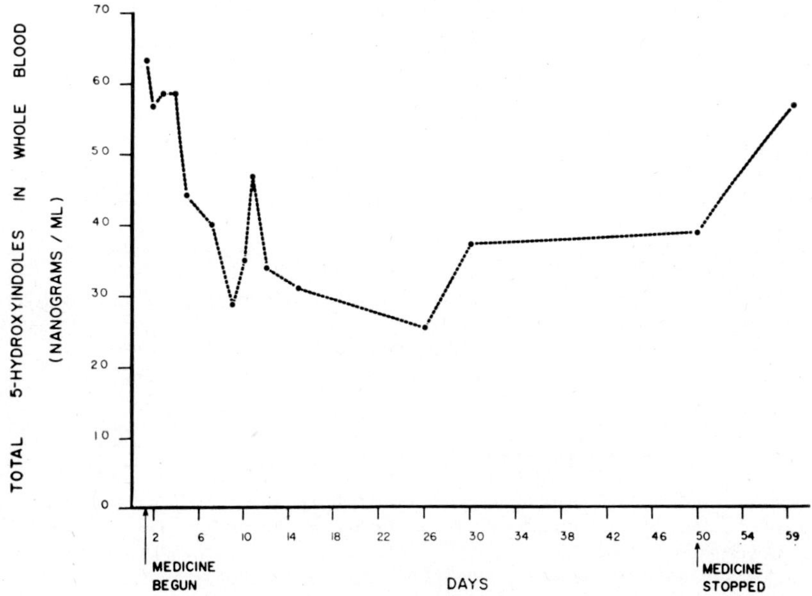

Fig. 8-4. A 9-year-old trisomy 21 patient, placed on imipramine HCl, had a depression of whole blood 5-HI levels while on the drug.

blockage of the platelet intracellular concentrating mechanism. There was no obvious clinical effect of the drug on this patient. No further trials are contemplated.

Conclusion

There are a number of alternative ways of elevating 5-HT levels in human beings. The methods considered in this chapter (MAOI, vitamin B6, tryptophan, tricyclic antidepressants) affect pathways other than the 5-HT pathway. Also, in our studies in Down's syndrome patients reported here, MAOI in a neonate and a tricyclic antidepressant in a 9-year-old patient caused a depression rather than elevation of whole blood 5-HI levels. The results of

MAOI in older children, vitamin B6 and tryptophan show very preliminary evidence of elevation of blood 5-HI levels beyond the expected placebo effect. These data remain to be confirmed by controlled studies.

Summary

Elevation of whole blood total 5-hydroxyindole levels was achieved in patients with Down's syndrome by the administration of vitamin B6 and tryptophan. Tranylcypromine, a monoamine oxidase inhibitor, lowered 5-HI levels in a neonate and elevated them in a 3- and a 9-year-old patient. Imipramine, a tricyclic antidepressant, lowered 5-HI levels in one 9-year-old patient.

REFERENCES

Beiler, J. M. (1954) Inhibition of 5-hydroxytryptophan decarboxylase. *J. Biol. Chem.* **211**, 39.

Benson, P. F. and Southgate, J. (1971) Diminished activity of platelet monoamine oxidase in Down's syndrome. *Am. J. Human Genetics*, **23**, 211.

Berman, J. L., Justice, P. and Hsia, D. Y. Y. (1965) The metabolism of 5-hydroxytryptamine in the newborn. *J. Pediat.* **67**, 603.

Bhagavan, H. N., Coleman, M., Coursin, D. B. and Rosenfeld, P. (1973) Pyridoxal-5-phosphate levels in the whole blood of home-reared patients with trisomy 21, *Lancet*, in press.

Bhagavan, H. N. and Coursin, D. B. (1971) Monosodium glutamate induces convulsive disorders in rats. *Nature* (*Lond.*), **232**, 275.

Bumbalo, T. S., Morelewicz, H. V. and Berens, D. L. (1964) Treatment of Down's syndrome with the "U" series of drugs. *J. Am. Med. Ass.* **187**, 361.

Carter, C. H. (1967) Unpredictability of mental development in Down's syndrome. *Southern Med. J.* **60**, 834.

Coburn, S. P. and Seidenberg, M. (1968) Leucocyte pyridoxal phosphate levels in Down's syndrome and other retardates. *Federation Proc.* **27**, 554.

Cohen, P. A., Schneidman, K., Ginsberg-Fellner, F., Sturman, J., Knittle, J. and Gaull, G. E. (1973) High pyridoxine diet in the rat. *J. Nutr.* **103**, 143.

Coleman, M. (1971) Infantile spasms associated with 5-hydroxytryptophan administration in patients with Down's syndrome. *Neurology*, **21**, 911.

Contractor, S. F. and Jeacock, M-K. (1967) A possible feed-back mechanism controlling the biosynthesis of 5-hydroxytryptamine. *Biochem. Pharmacol.* **16**, 1981.

Ebadi, M. S. (1970) Increase in brain pyridoxal phosphate by chlorpromazine. *Pharmacology*, **3**, 97.

Ebadi, M. S., Russell, R. L. and McCoy, E. E. (1968) The inverse relationship between the activity of pyridoxal kinase and the level of biogenic amines in rabbit brain. *J. Neurochem.* **15**, 659.

Fernstrom, J. D. and Wurtman, R. J. (1971) Brain serotonin content: physiological dependence on plasma tryptophan levels. *Science*, **173**, 149.

Finley, S. C., Finley, W. H., Rosecrans, C. J. and Phillips, C. (1965) Exceptional intelligence in a mongoloid child of a family with a 13–15/partial 21 (D/partial G) translocation. *New Engl. J. Med.* **272**, 1089.

Gallagher, B. B. (1971) The influence of tyrosine, phenylpyruvate and vitamin B6 upon seizure thresholds. *J. Neurochem.* **18**, 799.

GERSHOFF, S. N., HEGSTED, D. M. and TRULSON, M. F. (1958) Metabolic studies of mongoloids. *Amer. J. Clin. Nutr.* **6**, 526.

GOLDSTEIN, H. (1954) Treatment of mongolism and non-mongoloid mental retardation in children. *Arch. Pediat.* **11**, 77.

GOODMAN, H. O., KING, J. S. and THOMAS, J. J. (1964) Urinary excretion of beta-aminoisobutyric acid and taurine in mongolism. *Nature* (*Lond.*) **204**, 650.

HAUBOLD, H., LOEW, W. and HAEFELE-NIEMANN, R. (1960) Possibilities and limitations of a post-maturation treatment of retarded and especially of mongoloid children. *Landarzt*, **36**, 318.

HANSSON, O. (1969) Tryptophan loading and pyridoxine treatment in children with epilepsy. *Ann. N.Y. Acad. Sci.* **166**, 306.

HEATON-WARD, W. A. (1962) Inference and suggestion in a clinical trial (niamid in mongolism). *J. Ment. Sci.* **108**, 865.

JÉRÔME, H. (1962) Anomalies du métabolisme du tryptophane dans la maladie mongolienne. *Bull. Soc. Med. Hôp. Paris*, **113**, 168.

JÉRÔME, H., LEJEUNE, J. and TURPIN, R. (1960) Etude de l'excrétion urinaire de certains métabolites du tryptophane chez les enfants mongoliens. *C.R. Acad. Sci.* (*Paris*), **251**, 474.

KING, J. S., GOODMAN, H. O., WAINER, A. and THOMAS, J. J. (1968) Factors influencing urinary taurine excretion by normal and mongoloid subjects. *J. Nutrition*, **94**, 481.

KIRMAN, B. H. and PARE, C. M. (1961) Amine-oxidase inhibitors as possible treatment for PKU. *Lancet*, **1**, 117.

KIVALO, E., RINNE, U. K. and KARINKANTA, H. (1961) The effect of imipramine on the 5-hydroxytryptamine content and monoamine oxidase activity of the rat brain and on the excretion of 5-hydroxyindoleacetic acid. *J. Neurochem.* **8**, 105.

KONTSEVAYA, N. G. and RITSNER, M. C. (1972) Indices of tyrosine and tryptophan metabolism in children. *Pediatriia* (*Rus.*) **51**, 53.

KOPIN, I. J., FISCHER, J. E., MUSACCHIO, J. M., HORST, W. D. and WEISE, V. K. (1965) "False neurochemical transmitters" and the mechanism of sympathetic blockade by monoamine oxidase inhibitors. *J. Pharmacol. Exp. Ther.* **147**, 186.

LOVENBERG, W. (1970) Serotonin now: clinical implications of inhibiting its synthesis with parachlorophenylalanine. In: SJOERDSMA, A. *Ann. Int. Med.* **73**, 607.

MACON, J. B., SOKOLOFF, L. and GLOWINSKI, J. (1971) Feedback control of rat brain 5-hydroxytryptamine synthesis. *J. Neurochem.* **18**, 323.

MCCOY, E. E. and CHUNG, S. (1964) The excretion of tryptophan metabolites following deoxypyridoxine administration in mongoloid and non-mongoloid patients. *J. Pediat.* **64**, 227.

MCCOY, E. E., ROSTAFINSKY, M. J. and FISHBURN, C. (1968) The concentration of serotonin by platelets in Down's syndrome. *J. Ment. Defic. Res.* **12**, 18.

MCKEAN, C. The effect of high pyridoxine administration on the cerebral amine metabolism of infants with Down's syndrome, in preparation.

O'BRIEN, D. and GROSHEK, A. (1962) The abnormality of tryptophane metabolism in children with mongolism. *Arch. Dis. Child.* **37**, 17.

PAASONEN, M. K., SOLATUNTURI, E. and KIVALO, E. (1964) Monoamine oxidase activity of blood platelets and their ability to store 5-hydroxytryptamine in some mental deficiencies. *Psychopharmacologia*, **6**, 120.

PEREZ-CRUET, J., TAGLIAMONTE, A., TAGLIAMONTE, P. and GESSA, G. L. (1971) Stimulation of serotonin synthesis by lithium. *J. Pharmac. Exp. Ther.* **178**, 325.

PERRY, T. L. (1962) Urinary excretion of amines in phenylketonuria and mongolism. *Science*, **136**, 879.

RETT, A. (1959) Possibilities and limitation of therapy of brain injured children. *Wien. med. Wschr.* **108**, 1120.

ROBERGE, A. G., MISSALA, K. and SOURKES, T. L. (1971) Les systèmes serotoninergiques et le problème des faux transmetteurs. *Union Méd. Canada*, **100**, 475.

SCRIVER, C. R. and WHELAN, D. T. (1969) Glutamic acid decarboxylase (GAD) in mammalian tissue outside the central nervous system, and its possible relevance to hereditary vitamin B6 dependency with seizures. *Ann. N.Y. Acad. Sci.* **166**, 83.

SEGAWA, T. (1970) Effects of reserpine and desipramine on the uptake and subcellular distribution of 5-hydroxytryptamine in rabbit brainstem after intravenous administration of 5-hydroxytryptophan. *Jap. J. Pharmacol.* **20**, 87.

SHAW, M. W. (1962) Familial mongolism. *Cytogenetics*, **1**, 141.

SJOERDSMA, A., OATES, J. A., ZALTZMAN, P. and UDENFRIEND, S. (1959) Identification and assay of urinary tryptamine; application as an index of monoamine oxidase inhibition in man. *J. Pharmac. Exp. Ther.* **126**, 217.

SNELL, E. E. (1958) Chemical structure in relation to biological activities of vitamin B_6. *Vitam. and Horm.* **16**, 77.

SPECTOR, S., PROCKOP, D., SHORE, P. A. and BRODIE, B. B. (1958) Effect of iproniazid on the brain levels of norepinephrine and serotonin. *Science*, **127**, 704.

TANINO, Y. (1969) Improvement of children with mongolism through the effect of large dosage of various vitamins (2nd report). *Ann. Paediat. Jap.* **12**, 33.

TU, J. and ZELLWEGER, H. (1965) Blood serotonin deficiency in Down's syndrome. *Lancet* **2**, 715.

TURKEL, H. (1963) Medical treatment of mongolism. *Proc. 2nd Int. Cong. Ment. Retard.* **1**, 409.

VALADARES, J. R. E. (1966) The effects of pyridoxine deficiency on cardiac and brain phosphorylase in mice. *Biochim. Biophys. Acta*, **136**, 296.

VASQUEZ, H. J. and TURNER, M. (1961) Epilepsia en flexion generalizada. *Arch. Argent. Pediat.* **35**, 111.

WEISSBACH, H., BOGDANSKI, D. F., REDFIELD, B. G. and UDENFRIEND, S. (1957) Studies of the effect of vitamin B_6 on 5-hydroxytryptamine formation. *J. Biol. Chem.* **227**, 617.

WEISSBACH, H., LOVENBERG, W. and UDENFRIEND, S. (1961) Characteristics of mammalian histidine decarboxylating enzymes. *Biochim. Biophys. Acta*, **50**, 177.

WHITE, D. and KAPLITZ, S. E. (1964) Treatment of Down's syndrome with a vitamin–mineral–hormonal preparation. *Int. Copenhagen Cong. Ment. Retard.* **1**, 224.

WIECHERT, P. and GOLLNITZ, G. (1969) Stoffwechseluntersuchungen des cerebralen Anfallsgeschehens. *J. Neurochem.* **16**, 317.

CHAPTER 9

Platelet Serotonin in Disturbances of the Central Nervous System

MARY COLEMAN and FREDA HUR

The relevance of serotonin in brain disease is not limited to Down's syndrome. There are a large number of syndromes that affect 5-HT levels in blood (Erspamer, 1966; Warner, 1967; Page, 1968). Many of the disease entities involve the two major organ systems where 5-HT is synthesized; the central nervous system and the gastrointestinal tract. (For details on disease entities outside the brain, see Appendix IX-1.)

In the case of central nervous system disease, 5-HT abnormalities have been reported in 10 types of mental retardation and behavior disorders as well as in a number of patients without definite diagnoses (Table 9-1). In patients with

TABLE 9-1

CNS disease entities with serotonin abnormalities reported in blood

1. *Mental retardation*
   - Down's syndrome
   - Phenylketonuria
   - Histidinemia
   - Infantile spasm syndrome
   - Maternal rubella
   - Kernicterus
   - Infant hypothyroidism
   - Sudanophilic leukodystrophy
   - Non-specific syndromes—retarded patients with and without motor disabilities
   - deLange syndrome
2. *Behavior disorders*
   - Autistic syndromes
   - Hyperactivity syndromes

4 of these disease entities associated with mental retardation, adequately documented studies have been completed showing abnormal 5-HI blood levels

(Appendix IX-2). In addition to Down's syndrome, these are phenylketonuria (PKU), histidinemia and the infantile spasm syndrome. Also it appears that some patients with congenital organic brain syndromes affecting motor function ('cerebral palsy') have an abnormal 5-HT level. Also, many investigators have found 5-HT abnormalities in some retarded children without known specific diagnoses. It is reasonable to anticipate that the list of specific disease entities with 5-HT abnormalities will grow in the future (e.g. galactosemia; predicted by Woolley and Gommi, 1964).

One of the most intriguing aspects of 5-HT levels in blood is the close relationship of abnormal levels of 5-HT to decreased mental function in many of these patients. It is very apparent from Chapter 4 that 5-HT levels *per se* are unrelated to the development of intelligence in Down's syndrome patients. Yet a pattern of abnormal 5-HT blood levels/abnormal brain function and normal 5-HT blood levels/normal brain function is seen in several diseases. One example is PKU (most patients retarded) who have low 5-HT levels (Yarbro and Anderson, 1966); hyperphenylalaninemia patients (not retarded) have normal blood 5-HT levels (Berman *et al.*, 1969). According to Pare *et al.* (1960), spastics who are retarded may have abnormal serum 5-HT levels; spastic patients with normal intelligence quotients have normal levels.

Also, when patients with low 5-HT values are placed on diet therapies designed to improve brain function, the low 5-HT level will rise toward the normal range in the individual patient as he is treated. This has been reported in PKU (Pare *et al.*, 1958; Baldridge *et al.*, 1959) and histidinemia (Corner *et al.*, 1968). In a recent study of synthetic diets in retarded PKU patients, McKean (1971) reported a simultaneous "abrupt increase in blood serotonin and improved behavior" at phenylalanine levels of 10–12 mg/100 ml in 3 patients and 5 mg/100 ml in a fourth child. These critical levels of phenylalanine that improve behavior differ in each individual patient; it is most interesting that this is the same critical level that abruptly increases depressed blood serotonin levels. Conversely, when a hypothyroid patient is placed on thyroid therapy, his abnormally elevated whole blood 5-HI level drops into the normal range (Coleman, 1970).

Why is the platelet handling of 5-HT so useful clinically? Why does an uptake and binding system in an obscure blood cell appear to reflect abnormalities in CNS function? Paasonen (1968), Pletscher (1968) and Page (1968) have postulated that the platelet may be a partial 'model' for the serotonergic neuron. The answer probably will be found in the way 5-HT is regulated in the brain. Similar controls of synthesis in the brain and gastrointestinal tract (major source of platelet 5-HT) systems could be one answer. However, data can be interpreted to show that 5-HT synthesis in the brain is generally in excess of what is needed (Grahame-Smith, 1971; Greenberg and Coleman, 1973a). A more likely answer is found in the intraneuronal organization of

5-HT pools; that is, the system that critically determines the amount of 5-HT actually released at the receptor site in the serotonergic neuron. Under normal conditions, this is probably controlled by a complex system of ratios between unbound 5-HT, 5-HT bound to labile extravesicular binding sites and bound inside the storage vesicle (Carlsson, 1966; Takatsuka *et al.*, 1971). Ions and other factors that control binding may often control functional levels of 5-HT in the brain. These same factors may also control 5-HT binding to the amine granule in the platelet in many physiological states. This same process also appears to be highly sensitive to several types of pathological processes. Thus, the easily extracted platelet in the blood becomes a useful, if limited, tool in the study of the most inaccessible portion of the body—the central nervous system (Fig. 9-1).

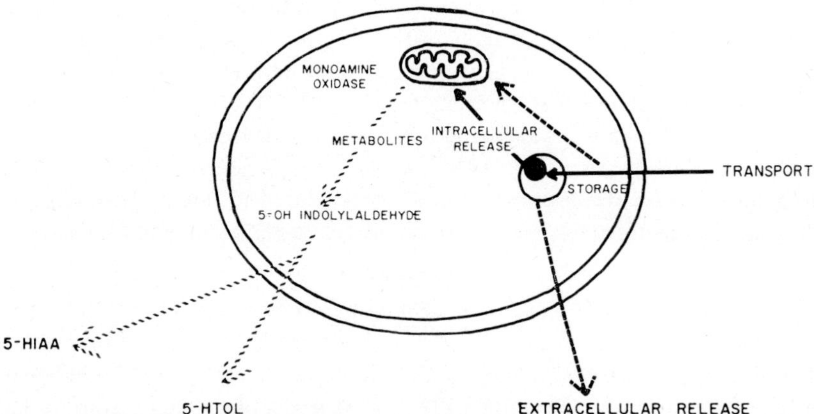

Fig. 9-1. The platelet 'model' for the serotonergic neuron. The platelet contains a particle that binds 5-HT. This binding site has some of the characteristics of the 5-HT binding site in neurons, thus making the platelet a partial 'model' for the serotonergic neuron (Paasonen, 1968; Pletscher, 1968; Page, 1968).

Based on this observation, a whole blood serotonin screening of normal newborns is underway in our laboratory as a first stage mental retardation screening test.

Specific Mental Retardation Syndromes where Serotonin may be Affected

Diseases where Two or More Authors Agree a Serotonin Abnormality Exists

Phenylketonuria

Low serotonin levels in serum or whole blood have been reported in classic phenylketonuria patients by several investigators (Pare *et al.*, 1957; Berendes *et al.*, 1958; Pare *et al.*, 1959; Yarbro and Anderson, 1966) and there is evidence from loading studies with radioactive tryptophan that the synthesis pathway is reduced by about 30% (Jequier, 1968). In our laboratory, we have also documented low whole blood levels in 3 patients (3, 38, and 44 ng/ml).

Loading high levels of phenylalanine into normal animals with intact phenylalanine hydroxylase enzyme systems gives a very imperfect model of PKU that does, however, include depression of central nervous system serotonin (Yuwiler and Louttit, 1961; Huang *et al.*, 1961; Boggs *et al.*, 1963). Limited clinical experimentation plus laboratory studies on these PKU model animals have been used to explore the various hypotheses of why high levels of phenylalanine or one of its metabolites may be related to lowered serotonin levels. There is some evidence for competitive inhibition of enzymes of the serotonin metabolic pathway—the rate limiting tryptophan hydroxylase seems the most likely of the enzymes to be depressed (Reichle *et al.*, 1961; Perry *et al.*, 1964; Lovenberg *et al.*, 1968). However, there is also a great deal of evidence that interference with active transport of the precursor amino acids may cause the serotonin depression (McKean *et al.*, 1962; Schanberg, 1963; McKean *et al.*, 1968; Yarbro and Anderson, 1966). Another hypothesis under study in our laboratory is related to failure of binding and storage of the amine, even if it is present in relatively normal amounts.

Histidinemia

Histidinemia occurs in children both retarded and with normal intelligence; it may be almost as prevalent as PKU. Low levels of platelet serotonin have been reported in one retarded patient (Corner *et al.*, 1968) and one child with a normal intelligence quotient but slow speech development (Auerbach *et al.*, 1962). Normal levels of 5-HT are recorded by Small and Holton (1970) in 2 children not retarded (Holton, personal communication). The etiology of the depressed serotonin in histidinemia is unknown; one possibility is that mechanisms similar to phenylketonuria are involved.

Infantile Spasms

The infantile spasm syndrome (with or without hypsarrhythmic EEG) is not a diagnostic entity. The symptomatic form of the syndrome is caused by almost any agent, such as metabolic disease, infections and subdurals that injure the infant central nervous system. However, there may be a common mechanism producing clinical symptoms in some of the patients since Ota (1969) reported elevated whole blood 5-HT in 7 out of 15 (46%) infantile spasm patients and we had confirmed this finding in 3 out of 6 (50%) patients

(Coleman *et al.*, 1971). We checked the blood 5-HI levels in this group of patients with spontaneous infantile spasms because of the 5-HTP side effect in this study (Chapter 7) and because the patients had clinical characteristics of the high serotonin syndromes (Table 9-2).

The tryptophan pathway has been involved in other previous studies in the infantile spasm syndrome. In 1958, Lowe and co-workers reported that depleting the diet of tryptophan resulted in clinical and EEG improvement in 1 out of 3 patients placed on such a diet (Lowe *et al.*, 1958). A year later, Cochrane found increased xanthurenic acid in the urine after a tryptophan loading test, which was considered evidence of a vitamin B6 deficiency (Cochrane, 1959). In addition to excessive xanthurenic acid, an abnormal 3-hydroxykynurenine to 3-hydroxyanthranilic acid ratio was found and French reported that giving B6 normalized these urine levels in a patient with infantile spasms associated with tuberous sclerosis (French, 1969). Bower (1961) and Cochrane also reported the use of B6 in the syndrome, but in spite of normalizing urine levels of these metabolites, no one was satisfied with any clinical improvement in the patients. Since then, several papers have reported normal plasma levels of pyridoxal phosphate in infantile spasms (Hagberg *et al.*, 1966; Yokoyama *et al.*, 1968). In some patients, B6 may actually be contraindicated because it also is the co-enzyme that forms 5-HT.

One problem of interpretation is that a seizure itself may change 5-HT levels in the central nervous system. Nellhaus has shown that there is an increase in 5-HT in the central nervous system of audiogenic seizure rabbits for several days following the clinical seizure (Nellhaus, 1968). However, most animal studies report that low 5-HT levels in the brain may be related to seizure susceptibility and that raising 5-HT helps prevent seizures (Nellhaus, 1968; Laborit *et al.*, 1958; de la Torre and Mullan, 1970).

Specific Diseases where One Author has Two or More Cases

Infant Hypothyroidism

Three patients with this disease entity have been studied in our laboratory prior to the initiation of replacement therapy. All 3 had abnormally high serotonin values (2 of these have been reported in the literature) (Coleman, 1970). Their values were: Patient A, 196 n/ml, Patient B, 192 g/ml and Patient C, 230 ng/ml. These levels are not as greatly elevated as are seen in the other high serotonin syndromes; however, they were repeatable in an elevated range for our laboratory. These results are the converse of the report of low serum 5-HT levels in 11 out of 19 patients with recent onset of hyperthyroidism and exopthalmos (Warner, 1967).

The mechanism of the serotonin abnormality in infants with hypothyroidism is unknown. Serotonin is related to thyroid function in two ways—the mesencephalic centers that regulate endocrine activity are affected by central nervous system levels and serotonin is present in the thyroid gland itself. Szanto influenced thyroid activity in rats by giving serotonin intragastrically (Szanto and Reviczky, 1965). Although his results showed an effect on both high and low levels of iodine-131 uptake by the thyroid, he concluded that serotonin characteristically inhibits thyroid activity by its role in de-iodination and hormone synthesis. Szanto also gave a serotonin inhibitor (methyl–lysergic acid–butanol–amide) and induced experimental hyperthyroidism. He then tried this inhibitor on human adults with myxedema and found that clinical symptoms and the usual tests of thyroid function improved. Conversely, Spencer has shown that high doses of thyroxine increased the toxicity of serotonin in mice (Spencer and Est, 1961).

Early initiation of thyroid replacement therapy in many patients with infant hypothyroidism appears of value in amelioration of the retardation seen in the untreated children (Raiti and Newns, 1971). We were able to follow 5-HI levels in 1 of our patients after therapy was initiated. In Patient A, the 5-HI level fell from 196 to 148 ng/ml 7 days after therapy was started and was 130 ng/ml 1 month after therapy began. Again, this is a pattern of 5-HI returning to the normal range after starting a treatment designed to improve central nervous system functioning, similar to the pattern seen in several of the low serotonin syndromes (PKU, histidinemia).

Maternal Rubella

In 1960, Pare, Sandler and Stacey made a survey of serotonin levels in 83 mentally defective patients drawn from institutions. They were the first (in 1957, 1959) to report high 5-HT levels in any retarded patients. Their 1960 paper showed that in 4 cases of maternal rubella there was an elevation of serum 5-HT above the normal range although in only 1 case was the urine 5-HIAA elevated.

Kernicterus

Pare and co-workers (1960) also studied 2 cases of kernicterus and reported that they both had high serum serotonin levels.

deLange Syndrome

Nine out of eleven patients with this syndrome have been reported to have low 5-HI levels in whole blood (Greenberg and Coleman, 1973b).

SINGLE CASE REPORTS IN THE LITERATURE

There are several individual case reports of 5-HT abnormalities in retarded patients (Appendix IX-2). In 2 cases where the reports are based on the serum method of determining 5-HT (familial dysautonomia) (Warner, 1967) and tuberous sclerosis (Pare *et al.*, 1960), we were unable to confirm an abnormality in other patients with these disease entities using the whole blood 5-HI method. We are currently preparing a single case report of a patient with sudanophilic leukodystrophy who had consistently elevated levels of whole blood 5-HI during his brief lifetime. Autopsy studies of 5-HT in major CNS nuclei disclosed a marked reduction of endogenous levels of 5-HT (Coleman *et al.*, in preparation).

Non-Specific Mental Retardation Syndromes

RETARDED PATIENTS WITH MOTOR DISABILITIES

Serum levels of 5-HT have been reported in institutionalized patients with congenital organic brain syndromes resulting in motor disabilities, popularly known as 'cerebral palsy'. According to Pare *et al.* (1960), 12 patients with average intelligence attending a residential school for spastics had normal levels of 5-HT while 28 institutionalized patients with I.Q.'s below 50 had elevated 5-HT levels. He states that high levels of 5-HT were equally present in retarded children with either spastic paraplegia or athetosis. Paasonen and Kivalo (1962) have confirmed high 5-HT levels in institutionalized patients, using both the serum and platelet methods. Tu and Partington (1972) have made a major contribution to the study of these patients by reporting normal levels of 5-HIAA in the CSF in spite of their high blood levels of 5-HT. We have studied 6 institutionalized, severely spastic patients and found the 1 most retarded had an elevated level, using the whole blood 5-HI method. The other 5 patients had 5-HI levels in or slightly below the normal range.

RETARDED PATIENTS WITHOUT MOTOR DISABILITIES

A number of authors have reported abnormal 5-HT levels in patients with types of mental retardation currently not classified. These patients are undoubtedly a mixed group of many different but unknown etiologies. Pare *et al.* (1960) studied 31 classified mental defectives who were retarded. They found that 20 had elevated serum 5-HT levels and 11 had normal levels. These patients were drawn from institutions. Paasonen and Kivalo (1962) also studied 5 patients who were severely retarded with the general classification of 'encephalopathia'. He reported they all had elevated serum 5-HT levels.

However, using his own platelet 5-HT levels on the same patients, one could interpret his results to indicate that 3 were elevated and 2 were normal. Schain and Freedman (1961) studied 7 'low grade' mental defectives and found their blood serotonin levels (estimated by bioassay) to be elevated. They compared them to middle and high grade defectives with normal levels and autistic patients with high levels. In 1965, Berman reported that 10 mentally retarded children, not otherwise described, had elevated whole blood 5-HT levels (Berman *et al.*, 1965).

We have studied 13 unclassified mentally retarded children, not institutionalized, and found that 10 of them had normal values, 1 had an abnormally low value and 2 were elevated. The patient with the low value (34 ng/ml, repeated 1 week apart with exactly the same value) was a black 4-year-old girl with cor pulmonale, microcephaly and mental retardation. Her brother's level of 5-HI was normal. Her severe congestive cardiac failure may have been a factor in her 5-HI value. One of the patients with high whole blood 5-HI levels (up to 600 ng/ml) has been studied extensively (Coleman and Barnet, 1970). We found her ATP levels in the platelets elevated. Pare *et al.* (1960) in a study of 4 other undefined patients with very high levels, found the ATP levels to be normal in his patients. Our patient was severely retarded and had no general or neurological abnormalities on examination except microcephaly and mild hypertonus. A sibling born recently in this family had a level of 171 ng/ml at 2 weeks of age and 272 ng/ml by 5 weeks of age. We also have a retarded patient, with an unknown diagnosis in spite of extensive investigation, with lower extremity hypertonus, nystagmus and failure of upward gaze who has 238 ng/ml of whole blood 5-HI.

Behavior Disorders

Autistic Syndromes

Schain and Freedman (1961) have described elevated levels of blood 5-HT in patients with 'autism'. A more detailed description of the patients is given in Schain and Yannett (1960) and they appear to be a group of withdrawn, disturbed children with a variety of neurological and psychiatric diagnoses. The series contains patients with hemiparesis, hydrocephalus and prematurity. (The end product of the 5-HT pathway—5-HIAA—is elevated in the cerebral spinal fluid of hydrocephalic patients; Rogers and Dubowitz, 1970.) Ritvo and co-authors (1969) have reported both increased 5-HT levels and increased platelet counts in a group of children "variously diagnosed as early infantile autism, atypical ego development, symbiotic psychosis and certain cases of childhood schizophrenia".

We have studied a number of patients with Kanner's primary autism and also retarded and psychotic children with autistic features. Although some patients do have elevated 5-HT levels, it is not consistently seen in any diagnostic category. A patient with Kanner's primary autism may have high, normal or low endogenous levels by the whole blood method. However, a primary autistic can be biochemically identified, in most cases, by a study of the 5-HT binding in the platelet. In our laboratory, we have found increased efflux of 5-HT from platelets of primary autistics, compared to controls and other types of psychotic children (Boullin *et al.*, 1970). In a recent study, the efflux ratio was used to predict primary autism from other psychiatric diagnoses, in a mixed group of disturbed children. There was a 90% correlation in this study between the efflux ratio prediction of autism and the clinical diagnosis of autism (Boullin *et al.*, 1971). This work has not yet been repeated by other laboratories. To date, the only other patient group other than Down's syndrome (Chapter 1) in our laboratory with a positive efflux ratio is the idiot savant group, patients with a clinical relationship to autism. The other study of platelet mechanisms of handling 5-HT in autistic patients was reported by Siva Sankar *et al.* (1963) who found reduced 5-HT uptake by platelets.

Hyperactivity Syndromes

We have recently described low 5-HI levels in 88% of a group of 25 children with a diagnosis of 'functional' hyperactivity (Coleman, 1971). In published studies on adult human beings, a relationship between lowered 5-hydroxyindole metabolism and clinical symptoms of depression has been postulated in some patients (Lapin and Oxenkrug, 1969; Bourne *et al.*, 1968; Van Praag and Korf, 1971a), although a study by Shaw *et al.* (1971) (which lacked controls) failed to show a change in binding in platelet 5-HT in depressed patients during and after their illness. Bunney and his colleagues have suggested there may be 2 types of clinical depression—a retarded, slowed-down patient with catecholamine deficit and an anxious, agitated patient with a serotonin deficit (Bunney *et al.*, 1969). Van Praag *et al.* (1971b) has also defined 2 types of depressed patients—those with and those without 5-HT deficit. It appears likely that some hyperactive children may belong to the agitated depressive category of patients with serotonin deficits.

Two of the hyperactive children with low 5-HI levels were placed in the hospital for detailed studies of their 5-HI metabolism. Their 5-HI levels rose while they were in a programmed hospital environment (Coleman, 1971). The elevation of 5-HI distinguishes this type of patient from normal children who have a lowering of whole blood 5-HI levels for the first few days after admission to a hospital. We have several cases that suggest that the 'paradoxical' response to hospitalization can even be used to predict future hyperactive children in studies on patients with unexplained low whole blood 5-HI levels under 18 months of age.

In a more recent study of institutionalized retardates with symptoms of the hyperactive syndrome, we have been able to show, by crossover studies, that blood 5-HT levels may be useful as a basis for selective drugs in the treatment of hyperactivity (Greenberg and Coleman, 1973).

Conclusion

In the first few weeks of life, a pattern of neurological examination that may correlate with high or low platelet serotonin levels has been noted in a number of these syndromes (Table 9-2). The main clinical difference between the 2 groups is in extremity tonus. It should be noted that *this clinical pattern is not limited only to patients with proven serotonin abnormalities*. We call these patient groups the serotonin syndromes because we happen to have a platelet model system where we can document abnormality in this amine in each patient. It is likely that other amines we have no method of measuring in a systemic 'model' system in humans are also consistently abnormal in these patients.

TABLE 9-2

Clinical characteristics of the serotonin syndromes of infancy

Clinical symptoms	Low platelet levels of 5-HT	High platelet levels of 5-HT
General		
Constipation	Present	Present
Neurological		
Mental retardation	Present	Present
Buccal–lingual hypotonia and dyskinesia	Present	Present
Neck traction response	Decreased	Decreased
Landau posture	Decreased	Variable
Tonus of extremities	Decreased	Increased

The high serotonin pattern can be seen in constipated, tonguing, infant patients who are hypothyroid. We studied a 4-month-old cretin who had marked head lag due to absent neck flexion in the traction response (undoubtedly why the literature reports these infants as 'floppy'). In spite of poor neck flexion (mediated primarily through cranial nerves), the patient had normal extensor tonus in the neck (spinal nerves) when held in the Landau posture (Fig. 9-2). He also had markedly *increased* tonus of all 4 extremities;

they were so hypertonic, it took marked effort to bend them. His 5-HI was 196 ng/ml the day this photograph was taken. A variation of this paradoxical tonus examination of neck hypotonia and extremity hypertonus has been described by Paine in another group now known to contain some patients with high serotonin levels—"infants who later developed spastic tetraparesis who are often hypotonic in the first year of life, especially in regard to the neck and trunk. Hyperreflexia, prolongation of the adductor, spread of the knee jerk after the normal age, and ankle clonus were early clues to future spasticity" (Paine *et al.*, 1964).

There is a 49-year-old patient in the literature who had neurological signs and high serotonin by a serum bioassay, but presumably, no mental retarda-

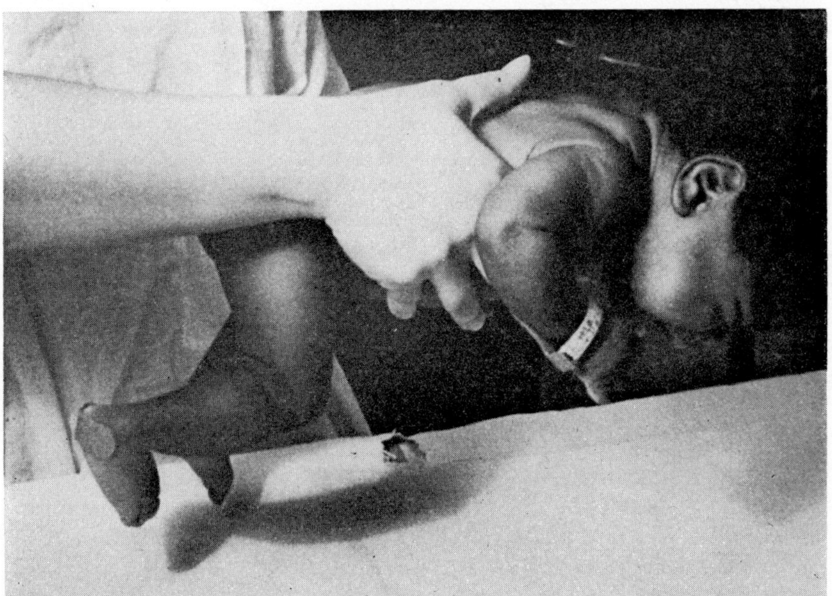

Fig. 9-2. This 4-month-old untreated patient with infant hypothyroidism has good extensor tone in the neck when held in the Landau posture.

tion (Southren *et al.*, 1959). (The authors were unable to determine whether she was taking the prescribed medicine so her mentation may not be totally intact; however, there is no mention of retardation in the case report.) Like many other high serotonin patients discussed in this chapter, she had an abnormal 5-HT level in blood (high) yet a normal 5-HIAA level in urine. No evidence of carcinoid tumor could be demonstrated, although she suffered from intermittent episodes of flushing of the head, neck and arms, hypertension and unresponsiveness. She also had a grand mal seizure; then experienced relief of all her episodes with diphenylhydantoin, a drug that, in addition to other known properties, does raise central nervous system 5-HT levels. Her neurological symptoms were "hand tremors, slurred speech, unsteadiness of gait and stiffness of her legs": in short, cerebellar signs and increased tonus. This case is particularly interesting since balance and tonus problems in retarded serotonin patients tend to resolve during childhood in many cases. (These clinical signs are also seen in 'occult' or 'normal pressure' adult hydrocephalus; in childhood hydrocephalus CSF levels of 5-HIAA are elevated. No measurement has been done in adult cases.) However, trials of 5-HTP and reserpine in this patient led the investigators to conclude that there was a "possibility that a functional deficiency of brain-tissue serotonin exists despite the high blood levels".

This study and our autopsy finding of low 5-HT in the brain of the Schilder's disease patient mentioned earlier suggests that the *high* blood serotonin patient group may have *low* central nervous system levels of the amine. It could be postulated that the high platelet levels are part of the body's feedback regulation system trying to compensate for *functional* low brain levels.

Perhaps both the high and low 5-HT levels in blood actually reflect low CNS levels in the mental retardation syndromes. Certainly all the syndromes have poor neck traction responses, an automatism controlled primarily by the brain (cranial nerve XI). The high platelet levels may be seen in diseases where brain 5-HT binding is not adequate; low platelet levels occur where an interference with synthesis by other aromatic amino acids (such as in PKU or histidinemia) prevents compensatory factors from operating. However, in Down's syndrome (Chapter 1) and the affective illnesses, this concept is clearly too simplistic.

In the low serotonin groups, the hypotonia of the extremities is well documented. Do these differences in extremity tone between the high and low 5-HT syndromes reflect differences in peripheral (in contrast to cerebral) factors? The hypotonia of Down's syndrome is unequivocal. Several authors (Lowe *et al.*, 1958; Paine and Fenichel, 1965) have also commented on hypotonia in untreated infant patients with phenylketonuria and histidinemia.

In addition to the patients with low 5-HT and hypotonia associated with

TABLE 9-3

Total 5-hydroxyindole levels in a family with benign congenital hypotonia

Patient	Age at time of 5-HI (years)	Sex	Age of walking (months)	Platelet count	5-HI (ng/ml)
A	6	Male	26	449,000	26
C	4	Male	20	400,000	32
B	3	Female	22	276,000	36

mental retardation, there is one interesting rare disease entity with hypotonia and low 5-HT where mental retardation is not present. Benign congenital hypotonia is an often-hoped-for, but rarely present, diagnostic entity. It is seen in families with many 'double jointed' members. We had the opportunity of studying whole blood levels of 5-HI on 3 children in such a family (Table 9-3) and found the levels to be quite low. No other known etiology of the low 5-HI levels could be identified in this family.

In conclusion, the information in the literature and in this monograph regarding serotonin in human beings suggest that the amine:

— is not directly related to brain systems mediating intelligence but is altered by many of the pathophysiological processes which alter these systems;
— is one of the mediators in those central nervous systems related to tonus and equilibrium;
— is one of the amines involved in the intricate balance regulating seizure thresholds and activity level;
— helps modulate sensory information reaching the central nervous system;
— affects the organization of sleep stages;
— is altered in acute states of depression and chronic affective disease;
— is one controlling factor in lower bowel peristalsis.

REFERENCES

AUERBACH, V. H., DIGEORGE, A. M., BALDRIDGE, R. C., TOURTELLOTE, C. D. and BRIGHAM, M. P. (1962) Histidinemia. *J. Pediat.* **60**, 487.

BALDRIDGE, R. C., BOROFSKY, L., BAIRD, H., REICHLE, F. and BULLOCK, D. (1959) Relationship of serum phenylalanine levels and ability of phenylketonurics to hydroxylate tryptophan. *Proc. Soc. Exp. Biol.* **100**, 529.

BERENDES, H., ANDERSON, J. A., ZIEGLER, M. B. and RUTTENBERG, D. (1958) Disturbance in tryptophan metabolism in phenylketonuria. *Amer. J. Dis. Child.* **96**, 430.

BERMAN, J. L., JUSTICE, P. and HSIA, D. Y. Y. (1965) The metabolism of 5-hydroxytryptamine (serotonin) in the newborn. *J. Pediat.* **67**, 603.

BERMAN, J. L., JUSTICE, P. and HSIA, D. Y. Y. (1969) Effect of vitamin B6 on blood 5-hydroxytryptamine concentration. *Ann. N.Y. Acad. Sci.* **166**, 97.

BOGGS, D. E., MCLAY, D., KAPPY, M. and WAISMAN, H. A. (1963) Excretion of indolyl acids in phenylketonuric monkeys. *Nature*, **200**, 76.

BOULLIN, D. J., COLEMAN, M. and O'BRIEN, R. A. (1970) Abnormalities in platelet 5-hydroxytryptamine efflux in patients with infantile autism. *Nature*, **226**, 371.

BOULLIN, D. J., COLEMAN, M., O'BRIEN, R. A. and RIMLAND, B. (1971) Laboratory predictions of infantile autism based on 5-hydroxytryptamine efflux from blood platelets and their correlation with the Rimland E-2 score. *J. Aut. Child. Schizophrenia*, **1**, 63.

BOURNE, H. R., BUNNEY, W. E., COLBURN, R. W., DAVIS, J. M., DAVIS, J. N., SHAW, D. M. and COPPEN, A. J. (1968) Noradrenalin, 5-hydroxytryptamine and 5-hydroxyindoleacetic acid in hindbrains of suicidal patients. *Lancet*, **2**, 805.

BOWER, B. D. (1961) The tryptophan load test in the syndrome of infantile spasms with oligophrenia. *Proc. Roy. Soc. Med.* **54**, 540.

BUNNEY, W. E., JANOWSKY, D. S., GOODWIN, F. K., DAVIS, J. M., BRODIE, H. K. H., MURPHY, D. L. and CHASE, T. N. (1969) Effect of L-DOPA on depression. *Lancet*, **1**, 885.

CARLSSON, A. (1966) In: Mechanisms of release of biogenic amines, Eds. Von Euller, U. S., Rosell, S. and Uvnas, B., Pergamon Press, New York.

COCHRANE, W. A. (1959) The syndrome of infantile spasms and progressive mental retardation related to amino acid and pyridoxine metabolism, International Congress of Pediatrics (Montreal), 8.

COLEMAN, M. (1970) Serotonin levels in infant hypothyroidism. *Lancet*, **2**, 365.

COLEMAN, M. (1971) Infantile spasms induced by 5-hydroxytryptophan in patients with Down's syndrome. *Neurology*, **21**, 911.

COLEMAN, M. and BARNET, A. (1970) Parachlorophenylalanine administration to a retarded patient with high blood serotonin levels. *Trans. Amer. Neurol. Ass.* **95**, 224.

COLEMAN, M., BOULLIN, D. J. and DAVIS, M. (1971) Serotonin abnormalities in the infantile spasm syndrome. *Neurology*, **21**, 421.

COLEMAN, M., LEE, J., HIJADA, D. and RANDALL, J. (1973) Low central nervous system and high blood levels of 5-hydroxytryptamine in a patient with sudanophilic leukodystrophy (in preparation).

CORNER, B. D., HALTON, J. B., NORMAN, R. M. and WILLIAMS, P. M. (1968) A case of histidinemia controlled with a low histidine diet. *Pediatrics*, **41**, 1074.

DE LA TORRE, J. C. and MULLAN, S. (1970) A possible role for 5-hydroxytryptamine in drug-induced seizures. *J. Pharm. Pharmac.* **22**, 858.

ERSPAMER, V. (1966) 5-Hydroxytryptamine and related indolealkylamines, Springer-Verlag, New York.

FRENCH, J. H. (1969) Pyridoxine and myoclonic seizure. *Ann. N.Y. Acad. Sci.* **166**, 310.

GRAHAME-SMITH, D. G. (1971) Studies *in vivo* on the relationship between brain tryptophan, brain 5-HT synthesis and hyperactivity in rats treated with a monoamine oxidase inhibitor and L-tryptophan. *J. Neurochem.* **18**, 1053.

GREENBERG, A. and COLEMAN, M. (1973a) Use of blood serotonin levels for the classification and treatment of hyperkinetic behavior disorders. *Neurology* (*Minneap.*), **23**, 428.

GREENBERG, A. and COLEMAN, M. (1973b) Depressed whole blood serotonin levels associated with behavioral abnormalities in the deLange syndrome. *Pediatrics*, in press.

HAGBERG, B., HAMFELT, A. and HANSSON, O. (1966) Tryptophan load test and pyridoxal-5-phosphate levels in epileptic children. *Acta Paediat.* **55**, 363.

HUANG, I., TANNENBAUM, S., BLUME, L. and HSIA, D. Y. Y. (1961) Metabolism of 5-hydroxyindole compounds in experimentally produced phenylketonuric rats. *Proc. Soc. Exp. Biol. Med.* **106**, 1961.

JEQUIER, E. (1968) Tryptophan hydroxylation in phenylketonuria. *Adv. Pharmac.* **6**, 169.

LABORIT, H., COIRAULT, R., BROUSSOLE, B., PERRIMOND-TROUCHET, R. and NIAUSSAT, P. (1958) Attempted interpretation of the mechanism of central action of various psychopharmacological agents (particularly serotonin) based on the results of an experimental convulsion induced by oxygen under pressure. *Ann. Médicopsychol.* (*Paris*), **116**, 60.

LAPIN, I. P. and OXENKRUG, G. F. (1969) Intensification of the central serotonergic processes as a possible determinant of the thymoleptic effect. *Lancet*, **1**, 132.

LOVENBERG, W., JEQUIER, E. and SJOERDSMA, A. (1968) Tryptophan hydroxylation in mammalian systems. *Adv. Pharmac.* **6**, 21.

LOWE, N. L., BOSMA, J. F., ARMSTRONG, M. D. and MADSEN, J. D. (1958) Infantile spasms with mental retardation: I. Clinical observations and dietary experiments. *Pediatrics*, **22**, 1153.

McKEAN, C. M. (1971) Effects of totally synthetic, low phenylalanine diet on adolescent phenylketonuric patients. *Arch. Dis. Child.* **46**, 608.

McKEAN, C. M., BOGGS, D. E. and PETERSON, N. A. (1968) The influence of high phenylalanine and tyrosine on the concentrations of essential amino acids in brain. *Chemistry*, **15**, 235.

McKEAN, C. M., SCHANBERG, S. M. and GIARMAN, N. J. (1962) A mechanism of the indole defect in experimental phenylketonuria. *Science*, **137**, 604.

NELLHAUS, G. (1968) Relationship of brain serotonin to convulsions. *Neurology*, **18**, 298.

OTA, S. (1969) Study of serotonin metabolism in pediatrics. 2. Blood serotonin levels in various diseases in children. *Acta Paediatrica Japonica*, **73**, 61.

PAASONEN, M. K. (1968) Platelet 5-hydroxytryptamine as a model in pharmacology. *Amer. Med. Exp. Biol. Fenn.* **46**, 416.

PAASONEN, M. K. and KIVALO, E. (1962) The inactivation of 5-hydroxytryptamine by blood platelets in mental deficiency with elevated serum 5-hydroxytryptamine. *Psychopharmacologia*, **3**, 188.

PAGE, I. H. (1968) Serotonin, Year Book Medical Publishers, Chicago.

PAINE, R., BRAZELTON, T. B., DONOVAN, D. E., DRORBAUGH, J. E., HUBBELL, J. P. and SEARS, E. M. (1964) Evolution of postural reflexes in normal infants and in the presence of chronic brain syndromes. *Neurology*, **14**, 1036.

PAINE, R. and FENICHEL, G. (1965) Infantile hypotonia. *Clin. Proc. Child. Hosp.* **21**, 175.

PARE, C. M. B., SANDLER, M. and STACEY, R. S. (1957) 5-Hydroxytryptamine deficiency in phenylketonuria. *Lancet*, **1**, 551.

PARE, C. M. B., SANDLER, M. and STACEY, R. S. (1958) Decreased 5-hydroxytryptophan decarboxylase activity in phenylketonuria. *Lancet*, **2**, 1099.

PARE, C. M. B., SANDLER, M. and STACEY, R. S. (1959) The relationship between decreased 5-hydroxyindole metabolism and mental defect in phenylketonuria. *Arch. Dis. Child.* **34**, 422.

PARE, C. M. B., SANDLER, M. and STACEY, R. S. (1960) 5-hydroxyindoles in mental deficiency. *J. Neurol. Neurosurg. Psychiat.* **23**, 341.

PERRY, T. L., HANSEN, S. and TISCHLER, B. (1964) Defective 5-hydroxylation of tryptophan in phenylketonuria. *Proc. Soc. Exp. Biol. Med.* **115**, 118.

PLETSCHER, A. (1968) Metabolism transfer and storage of 5-hydroxytryptamine in blood platelets. *Brit. J. Pharmac. Chemother.* **32**, 1.

RAITI, S. and NEWNS, G. H. (1971) Cretinism: Early diagnosis and its relation to mental prognosis. *Arch. Dis. Child.* **46**, 692.

REICHLE, F. A., BALDRIDGE, R. C., DOBBS, J. and TROMPETTER, M. (1961) Tryptophan metabolism in phenylketonuria. *J. Amer. Med. Ass.* **178**, 939.

RITVO, E. R., ORNITZ, E. M., EVIATAR, A., MARKHAM, C. H., BROWN, M. B. and MASON, A. (1969) Decreased postrotatory nystagmus in early infantile autism. *Neurology*, **19**, 653.

ROGERS, K. J. and DUBOWITZ, V. (1970) 5-Hydroxyindoles in hydrocephalus. A comparative study of cerebrospinal fluid and blood levels. *Develop. Med. Child. Neurol.* **12**, 461.

SCHAIN, R. J. and FREEDMAN, D. X. (1961) Studies on 5-hydroxyindole metabolism in autistic and other mentally retarded children. *J. Pediat.* **58**, 315.

SCHAIN, R. J. and YANNETT, H. (1960) Infantile autism. *J. Pediat.* **57**, 560.

SCHANBERG, S. M. (1963) A study of the transport of 5-hydroxytryptophan and 5-hydroxytryptamine (serotonin) into brain. *J. Pharmac. Exp. Ther.* **139**, 191.

SHAW, D. M., MACSWEENEY, D. A., WOOLCOCK, N. and BEVAN-JONE, A. B. (1971) Uptake and release of ^{14}C-5-hydroxytryptamine by platelets in affective illness. *J. Neurol. Neurosurg. Psychiat.* **34**, 224.

SIVA SANKAR, D. V., CATES, N., BROER, H. and SANKAR, D. B. (1963) Biochemical parameters of childhood schizophrenia (autism) and growth. In: J. Wortis (ed.) Recent advances in biochemical psychiatry, Plenum Press, **5**, 76.

SMALL, N. A. and HOLTON, J. B. (1970) Determination of platelet serotonin by a fluorimetric method. *Clinica Chimica Acta*, **27**, 171.

SOUTHREN, A. L., WARNER, R. R. P., CHRISTOFF, N. I. and WEINER, H. E. (1959) An unusual neurologic syndrome associated with hyperserotonemia. *New Engl. J. Med.* **260**, 1265.

SPENCER, P. S. J. and EST, G. B. (1961) Sensitivity of the hyperthyroid and hypothyroid mouse to histamine and 5-hydroxytryptamine. *Brit. J. Pharmac. Chemother.* **17**, 137.

SZANTO, L. and REVICZKY, A. L. (1965) Proc. Fifth Int. Thyroid Conf. (ed. C. Cassano and M. Andreoli), New York, 1164.

TAKATSUKA, K., SEWAGA, T. and TAKAGI, H. (1971) Uptake and storage mechanism of 5-hydroxytryptamine in rabbit brain stem and effect of reserpine. *Japan. J. Pharmacol.* **21**, 57.

TU, J. and PARTINGTON, M. (1972) 5-Hydroxyindole levels in the blood and CSF in Down's syndrome, phenylketonuria and severe mental retardation. *Develop. Med. Child Neurol.* **14**, 457.

VAN PRAAG, H. M. and KORF, J. (1971a) A pilot study of some kinetic aspects of the metabolism of 5-hydroxytryptamine in depressive patients. *Biol. Psychiat.* **3**, 105.

VAN PRAAG, H. M. and KORF, J. (1971b) Endogenous depressions with and without disturbances in the 5-hydroxytryptamine metabolism: a biochemical classification? *Psychopharmacologia*, **19**, 148.

WARNER, R. R. P. (1967) Current status and implications of serotonin in clinical medicine. *Adv. Int. Med.* **13**, 241.

WOOLLEY, D. W. and GOMMI, B. W. (1964) Serotonin receptors, IV. specific deficiency of receptors in galactose poisoning and its possible relationship to the idiocy of galactosemia. *Proc. Nat. Acad. Sci.* **52**, 14.

YARBRO, M. T. and ANDERSON, J. A. (1966) L-Tryptophan metabolism in phenylketonuria. *J. Pediat.* **68**, 895.

YOKOYAMA, Y., TADA, K., YOSHIDA, T., NAKAGAWA, H. and ARAKAWA, T. (1968) Liver kynureninase activity of an infant with infantile spasm. *Tohoku J. Exp. Med.* **96**, 191.

YUWILER, A. and LOUTTIT, R. T. (1961) Effects of phenylalanine diet on brain serotonin in the rat. *Science*, **134**, 831.

CHAPTER 10

Summary

Abnormalities in blood levels of serotonin have been reported in a number of types of mental retardation. This monograph explores the role of serotonin in Down's syndrome, a type of retardation associated with very low endogenous levels in the blood. A major portion of the book is devoted to a double blind, interdisciplinary evaluation of 19 patients with trisomy 21 who were given the precursor of serotonin, 5-hydroxytryptophan (5-HTP) or a placebo from the first week of life until 3 years of age.

Specific findings reported in this monograph are as follows.

A. Baseline Serotonin Studies

1. The infant age pattern of whole blood total 5-hydroxyindole (5-HI) levels was constructed based on samples from 174 normal children. In normal children, the mean 5-HI value under 24 hours of age was 73 nanograms/ml (which is 59% of their later childhood baseline of 124 nanograms/ml). The newborn value declined 15% to a new low between 36 and 48 hours of age; it then rose to achieve the childhood baseline between 15 and 21 days of age.

2. In 174 matching patients with trisomy 21, the mean 5-HI value under 24 hours of age was 34% lower than the value for normal children of the same age. The levels then rose more slowly than normal to a maximum level of 95 nanograms/ml at 2 months of age. Instead of maintaining this baseline as do normal children, the 5-HI level fell to another new low level of 49 nanograms/ml by 5 months of age. This low level then persisted as the childhood level for trisomy 21 patients; it is only 40% of the level of normal children.

During the first 2 months of life, patients with trisomy 21 mimic the 5-HI pattern of normal children at a lower level and a slower pace. After 2 months, the 5-HI value falls by 5 months to a new trisomy baseline maintained throughout childhood that is 40% of normal.

3. Twelve patients each with translocation and mosaic forms of Down's syndrome had low 5-HI levels not significantly different from trisomy 21

patients. Two patients with double chromosomal errors had 5-HI values higher and neurological examinations closer to normal than the trisomy 21 patients. 5-HI values in phenotypic mongols do not differ from those in normal children.

B. Double Blind 5-Hydroxytryptophan/Placebo Studies

1. *Mental function* The Bayley MDQ (Mental Development Quotient) in the patients receiving 5-HTP and those receiving placebo was not statistically different at 1, 2 or 3 years of age. This portion of the Bayley developmental scales appeared unaffected by 5-HTP administration.

2. *Motor function* (a) The Bayley PDQ (Psychomotor Development Quotient), which is primarily a measure of motor function, appeared to be affected by 5-HTP administration. At 12 and 24 months of age, the PDQ scores were significantly lower in patients receiving 5-HTP, although there was no statistically significant difference at 3 years of age.

(b) Neurological tonus scores also showed a 5-HTP effect. An increase in tonus was noted in newborns on the amino acid compared to placebo patients for the first 4 months of life. After that age, tonus scores were lower in patients receiving 5-HTP for the rest of the study until 3 years of age when they again became similar in both 5-HTP and placebo patients.

(c) These studies also revealed that a major factor affecting tonus in infants with Down's syndrome is environmental, particularly the 'attitudinal effect' of the mother. Hypotonus in young infants who were in an adverse environment could not be overcome with 5-HTP without raising doses to toxic levels.

(d) The earlier age of walking seen in the original 5-HTP studies in our clinic were not due to the effect of 5-HTP. In the double blind study, the same earlier age of walking was duplicated in placebo patients. Four out of 9 of the placebo patients walked by 16 months.

3. *EEG studies* 5-HTP appeared to exert a profound effect on the EEG parameters which were studied.

(a) The EEGs of 5-HTP treated patients showed a higher incidence of paroxysmal abnormalities than the placebo patients and the abnormalities were more severe. One patient in the double blind group developed seizures. They were of the infantile spasm type in this L-5-HTP treated patient and the EEG showed hypsarrythmia. Another D,L-5-HTP treated patient had a hypsarrythmic EEG but no clinical seizures. Other abnormalities included spiking and paroxysmal high voltage slow or mixed frequency bursts.

(b) Auditory evoked potentials tended to be of higher amplitude in 5-HTP patients than in placebo patients. Visual evoked responses were extremely variable for both groups but tended to show high amplitude and simpler form in 5-HTP patients compared to placebo patients. Sensory evoked potentials showed abnormal characteristics in both placebo and 5-HTP treated double blind patients.

(c) At 6 and 12 months of age, 5-HTP patients during afternoon sleep with sensory stimulation showed less stage REM than placebo patients. Both groups had little stage REM at 2 and 3 years of age. The amount and voltage of slow wave sleep was increased in 5-HTP patients over placebo patients.

(d) In the 3-year evaluation, abnormal EEGs were significantly associated with lower Bayley PDQ scores, but not with MDQ scores.

4. *Behavior patterns* (a) A significant positive correlation found at 6 months between pretreatment 5-HI levels and both MDQ and PDQ scores suggests that the status of this biochemical measure at birth possibly may be a predictor of early behavior development.

(b) There was a statistically significant difference between the 5-HTP and placebo groups on the Vineland Social Maturity scales at 2 years of age; the 5-HTP patients scored lower. This difference was absent by the 3-year examination.

(c) Psychiatric evaluation in their homes at 20 and 32 months of age showed no statistically different results on the 2 groups of patients. However, a trend toward lower scoring was seen in the 5-HTP patients. There was no statistically significant difference between the 2 groups in the scoring of the mother's attitude toward the patient.

(d) Compared to normal children, both the 5-HTP and placebo patients with Down's syndrome had a higher correlation between Bayley PDQ and MDQ scores.

(e) During the 3-year double blind study, abnormal behavior patterns such as head banging and self hitting occurred only in patients on placebo. Also, one 5-HTP patient began head banging when taken off the amino acid at the end of the study; reintroduction of 5-HTP coincided with a cessation of the head banging.

5. *Temperature studies* In the neonatal period, introduction of 5-HTP had no discernible effect on baseline temperature patterns.

6. *Strabismus* There was no difference between the 5-HTP and placebo patients in amount and severity of strabismus.

7. *Cardiac disease* 5-HTP had no discernible effect on the presence or severity of cardiac disease.

8. *Growth patterns* There was no statistical difference in height, weight or head sizes between the 5-HTP and placebo patients.

9. *Buccal–lingual hypotonia and dyskinesia* Despite the firm conviction of many observers that 5-HTP administration modified the buccal–lingual abnormalities of this syndrome, analysis of the 5-HTP/placebo data showed no difference between the 2 groups.

10. *Biochemical findings* (a) Administration of 5-HTP resulted in marked

elevation of whole blood 5-HI and urinary 5-HIAA levels. Whole blood total 5-hydroxyindoles were 57% to 99% 5-HT, 1% to 41% 5-HIAA and 0% to 13% 5-HTP in patients receiving large dosages of 5-HTP.

(b) A study of catecholamine metabolites disclosed a statistically significant elevation of epinephrine in the urine of patients chronically receiving 5-HTP compared to placebo patients. Other metabolites (norepinephrine, HVA, VMA, and MHPG) were not statistically different in the 2 groups.

11. *Side effects of 5-hydroxytryptophan* (a) A serious chronic effect of 5-HTP in Down's syndrome patients was the development of the infantile spasm syndrome in 14% of all patients who ever received the amino acid either in the double blind study or other clinic studies. Although in most cases it was reversible by lowering or stopping the dose of 5-HTP, ACTH was required to reverse the EEG and clinical symptoms in 1 case.

A clue to the role of serotonin in the spontaneous infantile spasm syndrome, another form of mental retardation, is provided by the 5-HTP toxicity study.

(b) In the double blind study, the EEG showed a higher incidence of background abnormalities in patients receiving 5-HTP than in the placebo patients and the abnormalities tended to be more pronounced in the 5-HTP group. At 3 years of age, a normal background EEG correlated with a higher Bayley PDQ (but not MDQ) score.

(c) Acute side effects of 5-HTP were diarrhea, hyperactivity, opisthotonus, hyperacousis and hypertension. Acute side effects were reversible by lowering the dose of 5-HTP. However, in ambulatory patients, too rapid a rate of withdrawal of 5-HTP was associated with ataxia.

C. Other Studies Affecting Serotonin Levels in Down's Syndrome

1. Whole blood 5-HI levels were raised in neonatal and 3-year-old patients with Down's syndrome by oral administration of vitamin B6 and tryptophan.

2. Tranylcypromine, a monoamine oxidase inhibitor, paradoxically lowered whole blood 5-HI levels to zero in a 2-week-old patient with trisomy 21. However, in 3- and 9-year-old patients, the MAOI elevated blood levels of 5-HI.

D. The Serotonin Syndromes of Infancy

The clinical pattern seen in patients with the serotonin syndromes of infancy is described. The patients have mental retardation, poor neck traction response, buccal–lingual dyskinesias and constipation. The serotonin syndromes include Down's syndrome as well as many other forms of retardation.

E. Based on the findings of this monograph, the administration of 5-HTP to patients with Down's syndrome is not recommended.

APPENDIX I

Appendix I-1

Total 5-hydroxyindole determinations

These methods are modifications of determinations discussed in Udenfriend, S., Weissbach, H. and Brodie, B. B. (1958) Methods of Biochemical Analysis, Volume VI.

WHOLE BLOOD MACRO METHOD

1. *Materials and Methods*

A plastic siliconized syringe is heparinized with 10,000 units per ml heparin. Care is taken to compress all heparin out of the syringe prior to venopuncture. 6–10 ml of blood is immediately emptied into tubes, IEC #1649, and *immediately* placed in dry ice. After 5–10 minutes it is placed in a laboratory freezer at $-10°C$. After 30 minutes in the freezer, it is placed into dry ice for 10 minutes and then placed in the freezer again for 10 minutes. Following this, the tube is placed in a glass homogenizer submerged in an ice bucket.

2. *Procedure*

2 ml of homogenized blood is pipetted into a tube containing 5.0 ml of cold non-fluorescent water, mixed immediately and placed in an ice bucket. 2.0 ml of 10% cold, non-fluorescent zinc sulfate is added, then shaken vigorously. Centrifuge in refrigerated centrifuge for 3 minutes at 16,000 rpm. Supernatant must be clear for re-spin. Carefully pipette 3.0 ml into tubes and add 0.9 ml 10N non-fluorescent HCl containing 0.1 mg% of EDTA, mix and read immediately. Three samples, one blank and one standard take about $2\frac{1}{2}$ hours. Read on a spectrofluorometer, Turner 110, at 295 nm (activation) and at 550 nm (fluorescence).

WHOLE BLOOD MICRO METHOD

1. *Materials and Methods*

Blood is collected by heel stick puncture in a neonatal patient into a plastic tube containing heparin. The method of drawing blood facilitates the collection. Approximately 1.2 ml blood is collected (for duplicate determinations) and immediately placed in dry ice. The blood is frozen and thawed three times as in the macro method.

2. *Procedure*

0.5 ml of whole blood is diluted with 1.0 ml of distilled, de-ionized water, mixed, then 0.5 ml of 10% $ZnSO_4$ is added, shaken on a shaker for 30 seconds, then 0.2 ml 1N NaOH is added and shaken on the shaker for 30 seconds. The mixture is then centrifuged in refrigerated centrifuge for 3 minutes at 16,000 rpm. The supernatant must be clear for re-spin. 0.1 ml HCl (non-fluorometric) is added to 1.0 ml of the clear supernatant. Read on spectrofluorometer, Turner 110, at 295 nm activation and 550 nm fluorescence. Run two samples, one blank and one standard simultaneously.

GENERAL COMMENTS

All determinations are run in duplicate with appropriate blanks and standards. It is important not to allow the blood to clot. If the heparinized step is omitted, we found an average 38 ng/ml loss of value.

Total 5-hydroxyindoles in whole blood decrease with time. Any blood not run immediately should be stored in a freezer not higher than −14°C. All determinations reported in this monograph were done within 4 days of venopuncture because we found levels began decreasing after that, even in freezers of −80°C.

Smoking and excessive dust in the laboratory may affect fluorescence.

Appendix I-2a

First visit total 5-hydroxyindoles in patients with trisomy 21

Age	Number of patients	Trisomy patient's 5-HI	Mean	Control's 5-HI	Mean	*P*
Less than 24 hr	5	27, 29, 38, 39, 71	41	54, 63, 78, 81, 90	73	0.016
1–2 days	13	5, 10, 30, 30, 39, 41, 44, 45, 46, 50, 50, 65, 80	44	33, 42, 54, 60, 66, 67, 69, 75, 78, 78, 81, 81, 81	67	0.001
2–3 days	12	14, 15, 20, 21, 22, 23, 27, 29, 30, 34, 36, 40	26	39, 45, 48, 48, 54, 57, 60, 60, 62, 69, 102, 102	62	<0.001
3–4 days	3	7, 25, 36	23	53, 68, 81	67	0.05
4–5 days	5	18, 20, 29, 34, 56	31	57, 62, 77, 92, 98	78 }	0.003
5–6 days	2	14, 31	23	75, 90	83 }	
6–7 days	10	5, 10, 15, 20, 20, 30, 33, 35, 54, 55	28	54, 65, 65, 66, 69, 78, 87, 93, 97, 103	78	<0.004
7–8 days	3	32, 39, 56	42	64, 66, 122	84	0.05
8–14 days	10	34, 38, 42, 46, 50, 50, 50, 50, 85, 87	53	63, 70, 72, 75, 87, 88, 93, 122, 160, 180	101	0.001
15–21 days	4	44, 45, 50, 59	50	112, 139, 145, 181	144	0.014
22–28 days	5	30, 36, 41, 76, 88	54	85, 98, 130, 151, 161	121	0.008
1 mth	8	44, 55, 56, 60, 67, 90, 91, 98	70	97, 103, 109, 130, 136, 140, 148, 172	129	<0.001
2 mth	7	39, 93, 99, 103, 105, 108, 116	95	89, 98, 108, 132, 147, 172, 180	132	0.07
3 mth	4	16, 49, 70, 81	54	90, 118, 122, 153	121	0.014
4 mth	6	36, 51, 72, 73, 91, 96	70	103, 108, 116, 135, 140, 157	127	0.001
5 mth	2	41, 82	62	92, 138	115	1.67
6 mth	5	34, 39, 46, 51, 80	50	107, 128, 148, 174	139	0.008
7 mth	1	50	50	130	130 }	
8 mth	1	55	55	163	163 }	
9 mth	0	—	—	—	— }	0.001
10 mth	2	54, 61	58	113, 142	128 }	
11–12 mth	3	26, 30, 88	48	104, 129, 134	122 }	
1–1½ yrs	7	26, 30, 36, 50, 60, 82, 83	64	98, 101, 118, 122, 125, 148, 156	124	0.001
1½–2 yrs	5	28, 40, 54, 57, 64	49	95, 95, 106, 135, 162	119	0.004

Appendix 1-2a (continued)

Age	Number of patients	Trisomy patient's 5-HI	Mean	Control's 5-HI	Mean	P
2–2½ yrs	7	22, 36, 44, 52, 66, 67, 87	53	90, 105, 106, 126, 126, 136, 170	123	0.001
2½–3 yrs	1	51	51	110	110	
3 yrs	6	28, 32, 35, 38, 39, 44	36	116, 116, 121, 131, 160	129	0.001
4 yrs	9	33, 33, 41, 52, 54, 54, 62, 75, 75	53	91, 108, 108, 110, 110, 114, 116, 145, 153	117	0.001
5 yrs	1	55	55	108	108	0.004
6 yrs	6	35, 37, 39, 45, 45, 66	45	98, 112, 119, 124, 134, 140, 148	125	
7 yrs	2	51, 68	60	121, 140	130	
8 yrs	3	61, 75, 80	72	106, 111, 168	128	0.001
9 yrs	3	46, 57, 75	56	102, 122, 168	131	
10 yrs	1	13	13	113	113	
11 yrs	0	—	—	—	—	
12 yrs	3	40, 44, 65	50	103, 119, 120	114	0.001
13 yrs	3	26, 42, 78	49	116, 127, 146	130	
14 yrs	0	—	—	—	—	
15 yrs	1	31	31	146	146	
16 yrs	1	29	29	164	164	
17 yrs	1	56	56	82	82	
18 yrs	0	—	—	—	—	
19 yrs	1	65	65	148	148	0.001
20 yrs	0	—	—	—	—	
21 yrs	0	—	—	—	—	
22 yrs	1	26	26	132	132	
26 yrs	1	54	54	94	94	

Appendix I-2b

First visit trisomy 21 patients—definition of the sample

Age when 5-HI was drawn		
	— under 30 days	— 64
	— 30 days to 1 year	— 37
	— over 1 year	— 72
Sex		
	male	— 93
	female	— 81
Race		
	Caucasian	— 148
	Black	— 26
Socio-economic classification		
	$3,000–$4,999	— 7
	$5,000– 6,999	— 16
	$7,000– 9,999	— 42
	$10,000–14,999	— 41
	$15,000–24,999	— 56
	$25,000 and over	— 12
Pelvic radiation in a parent		
	trisomy 21	— 59% (78 out of 132)
	mosaic	— 60% (3 out of 5)
	translocation	— 17% (1 out of 6)

Appendix I-3

First visit total 5-hydroxyindoles in patients with mosaic leukocyte karyotypes

Patient	Percentage of cells with 47 chromosome count	Age	Total 5-HI (ng/ml)	Compared to trisomy mean for age
C-1	50	10 yrs	150	+101
C-9	50	5 days	22	−5
C-10	51	1 yr 3 mth	48	−1
C-24	61	8 yrs	51	−7
C-11	65	4 mth	78	+8
C-7	75	5 mth	70	−13
C-4	75	9 days	50	+7
C-20	75	5 days	74	+47
C-22	80	1 yr 6 mth	85	+36
C-5	85	1 yr	35	−14
C-6	88	1 day	27	−14
C-13	90	2 mth	30	−65

Appendix 1-4

First visit total 5-hydroxyindoles in patients with translocation leukocyte karyotypes

Patient	Age	5-HI value	Compared to trisomy mean for age
G/D translocation			
C-8	New born	30	−11
C-18	3 days	65	+34
C-12	3 days	22	−9
C-19	3 days	31	0
C-14	9 days	46	−3
C-23	9 mth	30	−54
C-16	1 yr 3 mth	64	+15
Sum of difference between G/D mean and trisomy mean			−28
G/G translocation			
C-15	New born	18	−23
C-3	2 days	50	+24
C-17	3 days	70	+41
C-21	$3\frac{1}{2}$ mth	83	+13
C-2	4 yrs	37	−16
Sum of difference between G/G mean and trisomy mean			+39

Both types of translocation – grand total = +11.
Calculation of average increase: 11 ng ÷ 12 patients = <1 ng per patient.

Appendix 1-5

First visit 5-hydroxyindoles in phenotypical mongols

Patient	Chromosomes	Number of Down's syndrome stigmata	Cranial circumference	Intelligence quotient (Stanford-Binet)	First visit 5-HI ng/ml (whole blood)	Age
Pseudomongols						
G-1	46,XY	5	75th	56	126	4 yrs
G-7	46,XY	4	25th	42	103	5 yrs
Paramongols						
G-4	46,XX,21pss	7	50th	86	109	$2\frac{1}{2}$ yrs
G-6	46,XX	5	90th	89*	82	17 mth
G-2	46,XY	5	75th	107	52	7 mth
G-3	46,XX	5	45th	107	28	2 days
					55	4 days
					78	6 days
G-5	46,XY	5	25th	120	84	3 mth

* Slosson Intelligence Test at $2\frac{1}{2}$ years of age.

APPENDIX II

Appendix II-1

Patients on 5-HTP who had dose dropped rapidly rather than tapered

Patient	Age at time of drop	Reason for drop	Type of 5-HTP	Dosage of 5-HTP prior to drop	Dosage of 5-HTP after drop	5-HI levels prior to drop	5-HI levels after drop	Clinical effects of drop
A-8	4 yrs 10½ mth	Reoccurrence of seizure phenomena	L	4.4 mg b.i.d.	0	52	32	Clumsy, stumbled a lot, speech deterioration, back in 4 weeks, reappearance of strabismus
A-9	4½ yrs	Ran out of medicine	L	2.5 mg	0	103	80	Clumsiness, falling
B-8	3 yrs	To check effect on cardiac symptoms	D,L	36 cc/day	0	104	27	Off balance, falls when running, ? increase in lip cyanosis
B-10	3½ yrs	Intractable diarrhea	L	51.0 mg/day	20.0 mg/day	112	44	Ataxia, increased slurring and decreased amount of speech, increased sleeptime, increased buccal–lingual hypotonia
C-14	2 yrs	Persistent too high 5-HI level	D,L	24.0 mg/day	0	560	132	Ataxia after 3 days, falling, increased irritability
C-31	15 mth	Stopped for special study of 5-HI in this pt.	D,L	2.4 mg/kg	0	122	96	Less coordinated, was pulling to standing but lost ability to do so
D-9	2 yrs	Alopecia areata	D,L and B6	100 mg/day 50 mg/day	0	154	88	Ataxia after 2 wks, began stumbling, and falling a lot

Appendix II-2

Original group of trisomy 21 patients started on 5-HTP.
Age of walking and development quotients

Patient	Age of walking (mth)	MDQ (3 yrs)	MDQ (4 yrs)	PDQ (3 yrs)	PDQ (4 yrs)
A-1	18	47	47	45	39
A-2	15	52	51	73	62
A-3	25	51	62	49	60
A-4	15	64	53	54	58
A-5	18	63	60	51	42
A-6	40	38	—	24	—
A-7	24	56	42	44	50
A-8	23	64	63	47	50
A-9	21	—	—	—	—
A-10	17	40	—	51	—
A-11	22	53	—	63	—
A-12	23	—	—	—	—
A-13	17	62	—	46	—
A-14*	35	36**	—	15**	—

MDQ = Mental Development Quotient. PDQ = Psychomotor Development Quotient.
* = Institutionalized patient. ** = One year testing only.

Appendix II-3

Cardiac data on patients in this study (no Group A patients had cardiac symptoms)

Patient	Age	5-HI	Cardiac catheterization data: PA pressure (% systemic)	flow Qp/Qs	resistance PR/SR	Cardiac DX	Cardiac therapy: medical	surgical	Research therapy
B-4	19 mth	52	92%	1.9/1(5.4/1)	0.3(0.15)	VSD	0	0	Placebo
B-6	2 mth	90	100%	2/1	N.C.	ECD complete	+	PA banding	Placebo
B-8	4 mth	69	93%		0.9	ECD	+		D,L-5-HTP
	9 mth	150	89%		0.9	complete	+	PA	D,L-5-HTP
	11 mth	165	75%		0.55		+	banding	-0-
B-10	21 mth	96	51%	2.3/1	0.15	ECD complete	0	0	L-5-HTP
B-16	5 days	44 (7 days later)	70%	49/1	0.01	ECD complete	+	0	Placebo
B-18	10 mth	131	26%	1/1	0.1	aberrant R subclav.	0	Repair of vessel	L-5-HTP
C-3	—	—	—	—	—	VSD	+	0	Placebo
C-14	—	—	—	—	—	VSD	0	0	D,L-5-HTP, B6
D-10	10 mth	130	87%	2.2/1	0.3	VSD-PDA PABS	0	0	D,L-5-HTP, B6
E-13	—	—	—	—	—	ECD complete	0	0	B6
E-14	5 mth	56	72%	6.1/1	0.1	ECD complete	+	0	B6
E-16	—	—	—	—	—	ECD complete	+	0	B6
E-18	2 wks	102 (6 wks later)	100%	—	—	VSD	+	0	B6
E-26*	—	—	—	—	—	ECD, incom VSD	+		B6

VSD — ventricular septal defect
ECD — endocardial cushion defect
PDA — patent ductus arteriosus
PABS — pulmonary artery branch stenosis

Qp — pulmonary blood flow
Qs — systemic blood flow
PR — pulmonary resistance
SR — systemic resistance
5-HI — total 5-hydroxyindoles

L-5-HTP — 5-hydroxytryptophan, levo form
D,L-5-HTP — 5-hydroxytryptophan, racemic mixture
B6 — vitamin B6
* — deceased

APPENDIX III

Appendix III-1

Patients admitted, then withdrawn from double blind study

No.	Chromosomal karyotype	Age admitted	Age withdrawn	Reason for withdrawal from double blind study
1	47,XY,21+	7 days	5 wk	Deceased; bilateral lipid pneumonia
2	47,XX,21+	1 day	3 mth	Institutionalization
3	47,XX,21+	2 days	1 mth	Institutionalization
4	47,XX,21+	3 days	2 mth	Deceased; aspiration of gastric contents, cardiac A-V canal, pneumatosis intestinalis, Meckle's diverticulum
5	47,XY,21+	1 day	1 mth	Institutionalization
6	47,XY,21+	3 days	2 mth	Institutionalization
7	47,XX,21+/46,XX	2 days	2 mth	Mosaic karyotype
8	46,XY,D−,t(DgGg)+	1 day	18 mth	Translocation karyotype

Appendix III-2

Characteristics of patients in the double blind study (Group B)

Patient code	Sex	Race	Parental age father	Parental age mother	Abortions	No. of other siblings	Patients rank	Pre-conception pelvic X-ray in a parent	Socio-economic code
Placebo									
B-2	M	W	37	31	—	0	1	—	5
B-4	M	W	39	39	—	2	—	—	5
B-5	F	W	28	19	—	0	1	—	3
B-6	F	W	44	43	3	7	8	Yes	6
B-12	M	W	23	21	—	0	1	—	1
B-15	F	W	38	40	1	3	4	—	4
B-16	F	W	30–34	30–34	—	3	4	—	4
B-13	F	B	25	21	—	3	4	—	1
B-14	M	B	20	19	—	0	1	Yes	1
D,L-*5-HTP*									
B-1	F	W	30–34	40	—	6	5	—	4
B-3	F	W	40	36	—	0	1	—	5
B-7	F	W	38	40	—	8	9	Yes	3
B-8	F	W	37	34	2	1	2	—	4
B-11	F	W	43	43	1	4	5	Yes	3
L-*5-HTP*									
B-IS-19	F	W	42	42	1	0	1	—	4
B-10	F	W	55	37	—	0	1	Yes	1
B-9	M	W	23	19	—	0	1	Yes	4
B-17	F	W	39	33	2	1	2	—	4
B-18	M	W	35	35	—	—	—	—	5

Appendix III-3

Tonus scores—Group B patients

Patient	Pre-Rx				7 Days after starting Rx				14 Days		
	Age (days)	Tonus scores			Days after starting med.	Tonus score			Tonus score		
		truncal	U.E.	L.E.		truncal	U.E.	L.E.	truncal	U.E.	L.E.
B-2	4	3−	2+	2+	7	1	2	2	—	—	—
B-4	2	2	3	2	7	1+	2+	2−	—	—	—
B-5	—	3−	3−	3−	7	2	2+	2+	1+	2−	2−
B-6	—	1	1+	1+	7	1	1+	1+	2	2+	2+
B-12	2	1+	2	2	7	1+	2−	2−	1	1+	1+
B-13	—	—	—	—	—	—	—	—	—	—	—
B-14	1	1+	1+	1+	7	1	2−	2−	1+	2+	2+
B-15	3	2−	2	2	7	1	1	1	1	1	1
B-16	—	—	—	—	—	—	—	—	—	—	—
Placebo mean		1.9	2.0	1.9		1.1	1.7	1.7	1.2	1.6	1.6
B-1	—	2+	3−	3−	7	3	3	3	3	3	3
B-3	—	2	2−	2−	7	2	2−	2−	2−	3+	3+
B-7	—	0	1	1	—	1+	2	2	—	—	—
B-8	—	—	—	—	—	—	—	—	—	—	—
B-11	1	1	1+	1+	7	2−	1+	1+	3−	3−	3−
B-9	1	1+	2−	2−	7	2+	3−	3−	2	2	2
B-10	—	—	—	—	—	—	—	—	—	—	—
B-17	—	1−	1	1	7	1−	1	1	—	—	—
B-18	—	—	—	—	—	—	—	—	—	—	—
B-19	—	—	—	—	—	—	—	—	—	—	—
5-HTP mean		1.2	1.7	1.7		1.8	2.0	2.0	2.5	2.8	2.8

Patient	21 Days tonus scores			30 Days tonus scores			2 Months tonus scores			3 Months tonus scores		
	truncal	U.E.	L.E.	truncal	U.E.	L.E.	truncal	U.E.	L.E.	truncal	U.E.	L.E.
B-2	—	—	—	1+	2+	3−	0+	1+	1+	1+	2+	2
B-4	—	—	—	—	—	—	2	3+	3+	3	3	3
B-5	2−	2	2	3−	2+	2+	3	3	3	3	3	3
B-6	3−	3−	3−	1+	2	2	2	2	2	—	—	—
B-12	1+	2	2	1+	2+	2+	2+	2+	2+	2	2+	2+
B-13	—	—	—	4−	3	3	3	3	3	3	3−	3−
B-14	1	2	2	2	3−	3−	2−	2	2	3−	2+	2+
B-15	—	—	—	2	2+	2+	1+	2+	2+	1	3+	3+
B-16	—	—	—	1	2	2	—	—	—	1	2+	2+
Placebo mean	1.8	2.3	2.3	1.9	2.3	2.4	1.9	2.3	2.3	2.1	2.5	2.5
B-1	3	3−	3−	1	2	3	2	3	3	2+	3−	3−
B-3	—	—	—	1−	1	1	1	1	1	2	2	2
B-7	—	—	—	1+	2	2	4−	4−	4−	3	3	3
B-8	—	—	—	2	2−	2−	2	2	2	—	—	—
B-11	2	2	2	—	—	—	1	2	2	—	—	—
B-9	3	3	3	1+	3−	3−	3−	3	3	3−	3	3
B-10	—	—	—	3−	2+	2+	3	3−	3−	—	—	—
B-17	—	—	—	2+	3−	3−	2	4−	4−	—	—	—
B-18	—	—	—	1	2	2	2+	2+	2+	3	3	3−
B-19	—	—	—	3	3	3	3−	3−	3−	3	3	3
5-HTP mean	2.7	2.7	2.7	2.1	2.2	2.2	2.3	2.7	2.7	2.7	2.8	2.8

Patient	6 Months tonus scores			1 Year tonus scores			2 Years tonus scores			3 Years tonus scores	
	truncal	U.E.	L.E.	truncal	U.E.	L.E.	truncal	U.E.	L.E.	U.E.	L.E.
B-2	3–	2	2	3	3–	3–	3	2+	2+	2+	2+
B-4	2	2+	2+	1+	2	2	—	—	—	2+	2+
B-5	3	3	3	3	3	3	3	3	3	3	3
B-6	—	—	—	3	2–	2–	—	—	—	2+	3–
B-12	3	2+	2+	3	3–	3–	3	3	3	3	3
B-13	3	3–	2+	3	2+	2+	3	3–	3–	3	3
B-14	—	—	—	—	—	—	—	—	—	3–	3–
B-15	3–	2+	2+	3–	2	2	3	2	2	3–	3–
B-16	3–	2+	2–	2–	1+	1+	3–	2	2	2+	2+
Placebo mean	2.9	2.3	2.3	2.6	2.3	2.3	3.0	2.5	2.5	2.6	2.7
B-1	2	2	2	2*	2*	2*	2	2	2	3	3
B-3	1+	2	2	1+	1+	1+	3	3	3	3–	3–
B-7	—	—	—	2	2+	2+	3	3	3	3–*	3–*
B-8	2	2	2	2–**	2–**	2–**	3	2	2	3–	3–
B-11	—	—	—	3	3–	3–	—	—	—	3–	3–
B-9	—	—	—	3	3–	3–	3	2+	2+	3–	3–
B-10	3	2	2	3	3–	3–	3	1+*	1+*	3–	3–
B-17	2	1+	1+	2	2	2	3	2*	3–*	3–	1+
B-18	3	3	3	1	1	2	—	2	2	3–	2+
B-19	2+	2+	2+	2–	1+	1+	1	2+	2+	—	—
5-HTP mean	2.1	2.0	2.0	2.0	2.0	2.0	2.6	2.1	2.2	3.0	2.7

* 5-HI high. ** off 5-HTP.

Tonus scoring system

5	Severe hypertonus	2	Moderate hypotonia
4	Moderate hypertonus	2–	Moderate to severe hypotonia
3+	Mild hypertonus	1+	Severe hypotonia
3	Within normal limits	1	Very severe hypotonia
3–	Mild hypotonia	0	No tonus detected
2+	Mild to moderate hypotonia		

Appendix III-4

Group B patients,
total whole blood 5-HI, urinary 5-HIAA and 5-HTP dosages

Patient	1 Year			2 Years			3 Years		
	5-HTP (mg/kg)	5-HI	5-HIAA (μg/TV)	5-HTP (mg/kg)	5-HI	5-HIAA (μg/TV)	5-HTP (mg/kg)	5-HI	5-HIAA (μg/TV)
Placebo									
B-2	0	59	321	0	53	948	0	65	207
B-4	0	34	213	0	31	381	0	34	940
B-5	0	78	450	0	84	—	0	46	168
B-6	0	60	269	0	83	—	0	—	255
B-12	0	96	201	0	33	191	0	77	248
B-15	0	25	206	0	48	518	0	47	86
B-16	0	45	241	0	17	360	0	44	350
B-13	0	74	394	0	68	332	0	92	413
B-14	0	95	—	0	—	—	0	130	249
D,L									
B-1	41.2	169	13,040	27.2	383	8,453	9.7	120	10,850
B-3	10.5	98	9,636	10.8	141	12,502	16.6	196	13,877
B-7	13.1	81	4,970	24.0	105	5,779	34.3	224	21,105
B-8	14.2	127	939	21.2	112	5,751	17.1	104	5,510
B-11	10.1	95	3,313	—	—	—	—	—	—
L									
B-IS-19	0	78	188	4.8	117	—	12.8	96	11,505
B-10	13.8	126	3,229	19.4	342	5,950	12.0	250	8,700
B-9	14.8	224	8,823	18.7	276	23,000	11.0	204	9,750
B-17	9.7	132	2,600	19.1	320	15,635	11.4	108	9,490
B-18	11.0	139	1,148	18.1	344	6,216	6.8	280	4,185

Appendix III-5

Psychomotor milestones by EDC in Group B patients*

Patient	Age in months: Rolled over first time	Trans-ferred objects	Sat when placed	Walked alone	First English word	Parent report: Number of spontaneous, understand-able, English words at 3 years of age**	Number of word combina-tions
Placebo							
B-2	3/4	6	9	16	24	18	1
B-4	2	5	7	21	13	101	3
B-5	2½	4	6	14	13	30	5
B-6			13	33		10	0
B-12	3		7	23	15	42	2
B-13	3/4	3½	5	15	9	20	2
B-14	3		9	22			1
B-15	3	6½	11	16	9	103	53
B-16	6	8	17	36	20	2	0
Mean			9	22			
D,L							
B-1	3	6½	9	23	30	4	0
B-3	4	7	13	22	22	5	1
B-7	2½	5½	7	19	10	30	12
B-8	1		18	30	9	7	0
B-11	3½	6	10	17	13	Too many to count	A few small paragraphs
Mean			11	22			
L							
B-9	½	4	9	22	16	25	0
B-10	6	5½	12	21	18	94	4
B-17	3	7	16	31	15	2	0
B-18	6		19	27	34	2	0
B-19	3½	4	34	40	0	0	0
Mean (excluding B-19)			14	25			

* Obtained prospectively. ** Spoken words only; no singing words counted.

Appendix III-6a

Three-year 5-HIAA and catecholamine studies—Group B patients

Patient	5-HIAA ng/M²	Epi ng/M²	Nor ng/M²	HVA ng/M²	VMA ng/M²	MHPG ng/M²	MHPG ng/mg Cr
Placebo							
B-2	346	—	—	833	1,500	645	2.78
B-4	1,915	6.86	36.21	—	—	829	1.72
B-5	305	—	—	—	—	905	2.41
B-6	—	—	—	—	—	—	—
B-12	444	1.57	9.11	1,250	1,820	452	2.32
B-15	144	2.39	1.30	1,178	960	192	2.53
B-16	684	3.73	13.55	2,207	2,770	979	3.0
B-13	715	5.85	6.52	865	969	1,484	5.84
B-14	—	—	—	—	—	—	—
Mean	650	4.08	13.34	1,270	1,600	783.7	2.94
5-HTP							
B-1	—	—	—	—	—	—	1.34
B-3	21,888	5.84	8.68	946	2,330	651	2.72
B-7	37,354	11.10	10.90	1,590	2,600	625	1.71
B-8	5,510	16.20	21.98	1,840	4,100	646	2.52
B-IS-19	21,505	11.36	6.49	1,680	2,340	364	1.08
B-10	18,125	16.33	15.75	1,880	4,460	569	2.48
B-9	19,118	11.76	3.78	980	1,780	533	2.29
B-17	20,996	26.97	18.10	2,880	3,450	1,301	5.03
B-18	8,541	33.31	31.16	3,880	1,380	896	4.62
Mean	19,130	16.61	14.60	1,960	2,800	698.1	2.64
P	0.0005*	0.003*	**	**	0.047*	**	**

* Significant. ** Not significant.

Appendix III-6b

Three-year evaluations—catecholamine studies (cardiac patients omitted)

Patient	Epi (ng/M²)	Nor (ng/M²)	HVA (ng/M²)	VMA (ng/M²)
Placebo				
B-2	—	—	833	1,500
B-5	—	—	—	—
B-12	1.57	9.11	1,250	1,820
B-15	2.39	1.30	1,180	960
B-13	5.85	6.52	870	970
B-14	1.30	19.12	590	1,590
Mean	2.78	9.02	950	1,370
D,L-*5-HTP*				
B-1	—	—	—	—
B-3	5.84	8.68	946	2,330
B-7	11.10	10.90	1,593	2,600
B-11	—	—	—	—
Mean	8.47	9.79	1,270	2,470
L-*5-HTP*				
B-IS-19	11.36	6.49	1,680	2,340
B-9	11.76	3.78	980	1,780
B-17	26.97	18.10	2,880	3,450
Mean	16.70	9.46	1,850	2,530
Mean 5-HTP	13.41	9.59	1,620	2,500

Appendix III-7

Three-year evaluations—effect of 5-HTP on catecholamine metabolites (Group B patients)

Patient	dosage of 5-HTP Actual daily dose	Epi ratio	Nor ratio	HVA ratio	VMA ratio	MHPG ratio
D,L-*5-HTP*						
B-1	—	—	—	—	—	—
B-3	280.0	0.021	0.0310	0.0034	0.0083	2.325
B-7	470.0	0.024	0.0232	0.0034	0.0055	1.330
B-8	180.0	0.090	0.1221	0.0102	0.0228	3.589
Mean	310.0	0.0463	0.0588	0.0057	0.0122	2.415
L-*5-HTP*						
B-IS-19	13.8	0.823	0.4703	0.1217	0.1696	26.377
B-10	250.0	0.065	0.0630	0.0075	0.0178	2.276
B-9	192.0	0.061	0.0197	0.0051	0.0093	2.776
B-17	217.5	0.124	0.0832	0.0133	0.0159	5.982
B-18	90.0	0.370	0.3462	0.0431	0.0153	9.956
Mean	152.7	0.2886	0.1965	0.0381	0.0456	9.473
P	**	0.071*	**	**	**	**

* Not quite significant. ** Not significant.

Appendix III-8

Height, weight and cranial circumferences in double blind patients three years of age

Patient code	Height (cm)	Weight (kg)	Cranial circumference (cm)
Placebo			
B-2	89.5	14.5	48.0
B-4	83.8	10.9	47.0
B-5	91.4	12.2	46.0
B-6	90.2	15.0	47.0
B-12	88.4	13.8	47.0
B-15	92.7	14.7	47.0
B-16	84.5	11.8	44.0
B-13	—	—	45.5
B-14	83.8	13.4	47.5
Mean	88.0	13.3	46.6
D,L-5-*HTP*			
B-1	—	10.3	46.75
B-3	90.2	16.9	48.0
B-7	88.9	13.7	48.0
B-8	85.4	10.5	45.0
B-11	—	—	46.25
Mean	88.2	12.8	46.8
L-5-*HTP*			
B-IS-19	90.2	11.7	44.0
B-10	83.8	10.4	44.25
B-9	89.0	10.9	48.0
B-17	80.6	9.6	43.5
B-18	86.4	10.2	46.5
Mean	86.0	10.6	45.25
Mean D,L- & L-5-HTP	86.8	11.6	46.25

Appendix III-9

Three-year evaluations—platelet count (Group B patients)

Placebo patients	*Platelet count*
B-2	190,722
B-4	94,024
B-5	212,289
B-6	
B-12	173,097
B-15	160,680
B-16	152,342
B-13	207,000
B-14	
Mean	170,022
D,L-*5-HTP patients*	
B-1	210,022
B-3	245,590
B-7	155,509
B-8	—
Mean	203,707
L-*5-HTP patients*	
B-IS-19	173,118
B-10	301,399
B-9	301,080
B-17	94,491
B-18	164,522
Mean	206,922
Mean combined 5-HTP	205,716

The platelet count is not significantly increased (using either the *t*-test or the two sample rank sum test).

Appendix IV-1

Comparison of mean Bayley Mental Development Quotients* (MDQ) of Down's syndrome infants in 5-HTP treated and placebo groups during the first three years of life

Chronological age		Placebo group	5-HTP treated group			t Ratios					
			D,L subgroup	L subgroup	Total 5-HTP group	Placebo *vs.* D,L subgroup		Placebo *vs.* L subgroup		Placebo *vs.* total 5-HTP group	
						t	*P*(2-tailed)	*t*	*P*(2-tailed)	*t*	*P*(2-tailed)
	n	8	5	3	8						
3 mth	XMDQ	106.9	85.6	91.7	87.9	1.180	<0.30	0.662	>0.50	1.327	<0.30
	SD	37.7	16.3	14.4	14.9						
	n	8	4	3	7						
6 mth	XMDQ	90.9	82.2	87.3	84.4	1.030	<0.30	0.350	>0.50	0.926	<0.40
	SD	15.0	9.9	14.8	11.4						
	n	8	5	5	10						
1 yr	XMDQ	69.8	66.2	58.4	62.3	0.423	>0.50	1.161	<0.30	0.993	<0.40
	SD	15.9	12.4	19.2	15.7						
	n	7	4	3	7						
2 yrs	XMDQ	58.3	58.0	58.0	58.0	0.046	>0.50	0.044	>0.50	0.148	>0.50
	SD	10.4	9.1	5.3	7.1						
	n	9	4	5	9						
3 yrs	XMDQ	52.1	52.5	46.8	49.3	0.115	>0.50	0.926	<0.40	0.639	>0.50
	SD	6.4	2.6	15.3	11.4						

* Because the usual method of reporting Bayley Scale results in terms of deviation indices (MDI and PDI) cannot be meaningfully applied when these scores fall below 50, or with subjects beyond the age of 30 months. Ratio Quotients resulting from dividing a subject's Mental or Motor Age Equivalent by his Chronological Age (× 100) were used throughout this study.

Appendix IV-2

Comparison of mean Bayley Psychomotor Development Quotients* (PDQ) of Down's syndrome infants in 5-HTP treated and placebo groups during the first three years of life

Chrono-logical age		Placebo group	5-HTP treated group			*t* Ratios					
			D,L subgroup	L subgroup	Total 5-HTP group	Placebo *vs.* D,L subgroup		Placebo *vs.* L subgroup		Placebo *vs.* total 5-HTP group	
						t	*P*(2-tailed)	*t*	*P*(2-tailed)	*t*	*P*(2-tailed)
	n	8	5	3	8						
3 mth	XPDQ	115.5	76.8	80.7	78.2	2.536	< 0.05	1.769	< 0.20	3.023	< 0.01
	SD	31.7	14.7	17.2	14.6						
	n	8	4	3	7						
6 mth	XPDQ	88.2	81.2	80.0	80.7	0.845	< 0.50	0.812	> 0.40	1.025	< 0.40
	SD	13.3	14.1	19.9	15.2						
	n	8	5	5	10						
1 yr	XPDQ	62.8	55.8	45.4	50.6	0.958	< 0.40	2.950	< 0.02	1.972	< 0.10
	SD	12.7	12.8	12.7	13.2						
	n	7	4	3	7						
2 yrs	XPDQ	61.0	50.0	41.3	46.3	1.369	< 0.30	2.756	< 0.05	2.325	< 0.05
	SD	11.4	15.2	6.1	12.3						
	n	9	4	5	9						
3 yrs	XPDQ	54.8	53.2	40.6	46.2	0.276	> 0.50	2.165	< 0.10	1.585	< 0.20
	SD	10.3	5.4	14.2	12.5						

* Because the usual method of reporting Bayley Scale results in terms of deviation indices (MDI and PDI) cannot be meaningfully applied when these scores fall below 50, or with subjects beyond the age of 30 months. Ratio Quotients resulting from dividing a subject's Mental or Motor Age Equivalent by his Chronological Age (× 100) were used throughout this study.

Appendix IV-3

Selected Mental and Motor Bayley Scale items in various behavior categories: percentages passed by total Down's syndrome subject group by age

Behavior category*	3 Months ($n=16$)			6 Months ($n=15$)		
	Item	$\bar{x}$ Age normal**	% DS group passing	Item	$\bar{x}$ Age normal	% DS group passing
Sensory responsiveness	Head follows dangling ring	3.2	56	Regards pellet	4.7	73
	Turns head to sound of bell	3.8	12	Interest in sound production	5.8	33
Gross motor	Holds head steady	2.5	31	Head balanced	4.2	87
				Turns from back to side	4.4	53
	Sits with slight support	3.8	75			
				Pulls to sitting	5.3	20
				Sits alone momentarily	5.3	7
Fine motor	Fingers hand in play	3.2	44	Cube: partial thumb opposition (radial–palmar)	4.9	40
				Transfers object hand to hand	5.5	33
	Cube: ulnar palmar prehension	3.7	25	Scoops pellet	6.8	0
Motivation: exploration + purposiveness	Reaches for dangling ring	3.1	88	Reaches persistently	5.0	27
				Sustained inspection of ring	5.4	33
				Bangs in play	5.6	60
Social, gestural and imitative	Social smile: E smiles, quiet	2.1	50	Discriminates strangers	4.8	87
				Likes frolic play	5.1	13
				Smiles at mirror image	5.4	47
Perceptuocognitive	Aware of strange situation	3.5	44	Turns head after fallen spoon	5.2	67
				Pulls string: secures ring	5.7	27
				Uncovers toy	8.1	0
Language and vocalization	Vocalizes 2 different sounds	2.3	44	Vocalizes attitudes	4.6	87
				Vocalizes 4 different syllables	7.0	0

* From Bayley Infant Scale Performance Profile developed by A. Lodge. ** Bayley (1969).

Appendix IV-4

Selected Mental and Motor Bayley Scale items in various behavior categories: percentages passed by total Down's syndrome subject group by age

Behavior category*	1 Year			2 Years			3 Years		
	Item	$\bar{x}$ Age normal**	% DS group passing	Item	$\bar{x}$ Age normal	% DS group passing	Item	$\bar{x}$ Age normal	% DS group passing
Gross motor	Sits alone, good coordination	6.9	59	Walks alone	11.7	57	Tries to stand on walking board	17.8	72
	Prewalking progression	7.1	65	Throws ball	13.3	93	Stands up: II	21.9	44
	Stands alone	11.0	6	Walks up stairs with help	16.1	29	Walks on line, general direction	23.9	28
							Walks up stairs alone: both feet on each step	25.1	6
Fine motor	Cube: complete thumb opposition (radial–digital)	6.9	41	Pellet: fine prehension (neat pincer)	8.9	71	Pegs placed in 70 seconds	16.4	61
	Pellet: partial finger prehension (inferior pincer)	7.4	12	Holds crayon adaptively	11.2	93	Builds tower of 3 cubes	16.7	72
	Pat-a-cake: midline skill	9.7	35	Places 1 peg repeatedly	13.0	43	Folds paper	27.9	11
Motivation: exploration + pur-posiveness	Attempts to secure 3 cubes	7.7	47	3 or more cubes in cup	12.4	100	Blue board: places 2 round + 2 square blocks	19.3	50
	Fingers holes in peg board	8.9	29	9 cubes in cup	13.7	36			
Social, gestural and imitative	Cooperates in games	7.6	82	Pats whistle doll, in imitation	12.2	86	Imitates crayon stroke	17.8	56
	Stirs with spoon in imitation	9.7	35	Uses gestures to make wants known	14.6	86	Imitates strokes: vertical + horizontal	24.4	22
	Repeats performance laughed at	10.8	29						
Perceptuo-cognitive	Picks up cup: secures cube	9.0	24	Blue board: places 1 round block	13.6	64	Finds 2 hidden objects	19.7	50
	Unwraps cube	10.5	18	Pink board: places round block	16.8	36	Pink board: completes watch	21.2	44
	Removes pellet from bottle	13.4	0	Attains toy with stick	17.0	43	Names 4th picture	23.8	0
Language and vocal-ization	Says "da-da" or equivalent	7.9	76	Says 2 words	14.2	50	Uses words to make wants known	18.8	56
	Inhibits on command	10.1	41	Names 1 object	17.8	14	Points to parts of doll	19.1	56
	Jabbers expressively	12.0	0	Follows directions: doll	17.8	29	Names 1 picture	19.3	44
							Sentence of 2 words	20.6	17

* From Bayley Infant Scale Performance Profile developed by A. Lodge. ** Bayley (1969).

Appendix IV-5

Comparison of mean Vineland Social Quotients (SQ) of Down's syndrome infants in 5-HTP treated and placebo groups during the first three years of life

Chronological age		Placebo group	5-HTP treated group			*t* Ratios					
			D,L subgroup	L subgroup	Total 5-HTP group	Placebo *vs.* D,L subgroup		Placebo *vs.* L subgroup		Placebo *vs.* total 5-HTP group	
						t	*P*(2-tailed)	*t*	*P*(2-tailed)	*t*	*P*(2-tailed)
	n	8	4	5	9						
1 yr	XSQ	74.6	68.5	54.4	60.7	0.615	>0.50	1.796	<0.10	1.753	<0.20
	SD	17.3	13.3	18.7	17.2						
	R	47–89	55–85	22–68	22–85						
	n	7	4	3	7						
2 yrs	XSQ	77.4	64.0	64.3	64.1	2.024	<0.10	2.043	<0.10	2.510	<0.05
	SD	9.3	12.7	9.3	10.5						
	R	61–87	49–80	58–75	49–80						
	n	8	4	5	9						
3 yrs	XSQ	73.8	72.8	57.2	64.1	0.137	>0.50	1.928	<0.10	1.389	<0.20
	SD	14.1	3.1	16.5	14.4						
	R	51–90	70–77	34–78	34–78						

Appendix IV-6

Mean Mental, Motor and Social Age Equivalents (in months) of Down's syndrome infants in 5-HTP treated and placebo groups during the first three years of life

	Placebo group				5-HTP treated group: D,L subgroup				5-HTP treated group: L-subgroup				Total 5-HTP group			
	Chronological	Mental	Motor	Social	Chronological	Mental	Motor	Social	Chronological	Mental	Motor	Social	Chronological	Mental	Motor	Social
	(ages in months)				(ages in months)				(ages in months)				(ages in months)			
	$n=8$				$n=5$				$n=3$				$n=8$			
X=	3.2	3.3	3.6		3.2	2.8	2.5		3.3	3.0	2.7		3.2	2.9	2.6	
SD=	0.5	1.1	0.9		0.9	1.0	0.9		0.6	0.0	0.6		0.5	0.7	0.7	
R=	2–4	2–5	2–4		3–4	2–4	2–4		3–4	3–3	2–3		3–4	2–4	2–4	
	$n=8$				$n=4$				$n=3$				$n=7$			
X=	6.0	5.4	5.2		6.0	4.9	4.9		6.2	5.3	4.8		6.1	5.1	4.9	
SD=	0.5	0.7	0.6		–0–	0.6	0.8		0.8	0.6	0.8		1.4	0.6	0.7	
R=	5–7	4–7	4–6		6–6	4–6	4–6		6–7	5–6	4–6		6–7	4–6	4–6	
	$n=8$				$n=5$				$n=5$				$n=10$			
X=	12.8	8.9	8.0	9.5	12.8	8.4	7.1	8.8	12.6	7.3	5.7	6.9	12.7	7.8	6.4	7.8
SD=	0.7	2.1	1.7	2.3	0.8	1.1	1.4	1.3	0.6	2.3	1.5	2.4	0.7	1.8	1.6	2.16
R=	12–14	6–11	6–10	6–12	12–14	7–10	6–9	7–10	12–13	4–9	4–7	3–10	12–14	4–10	4–9	3–10
	$n=7$				$n=4$				$n=3$				$n=7$			
X=	25.8	15.0	15.9	20.0	24.8	14.2	12.5	15.9	25.0	14.5	10.3	16.0	24.9	14.4	11.6	16.0
SD=	1.3	2.5	3.9	2.6	1.3	1.7	4.2	2.2	–0–	1.3	1.5	2.3	0.9	1.4	3.3	2.1
R=	24–28	12–18	8–20	15–23	23–26	12–16	8–17	13–18	25–25	14–16	9–12	14–19	23–26	12–16	8–17	13–19
	$n=9$				$n=4$				$n=5$				$n=9$			
X=	37.1	19.2	20.2	27.2	36.2	19.0	19.2	26.4	37.0	17.2	14.8	21.1	36.7	18.0	16.8	23.4
SD=	2.0	1.9	3.4	4.3	0.5	0.8	1.9	1.0	1.2	5.2	5.0	5.8	1.0	3.9	4.4	5.0
R=	36–42	15–22	14–24	21–32	36–37	18–20	18–22	25–28	36–39	8–21	8–19	13–29	36–39	8–21	8–22	13–29

Appendix IV-7

Vineland Social Maturity Scale results at two years of age: mean point scores in various behavior categories obtained by 5-HTP treated and placebo infant groups

Behavior category	Placebo group	5-HTP Group			Comparison of placebo *vs.* total 5-HTP group
		D, L subgroup	L subgroup	Total 5-HTP group	
	$n=7$	$n=4$	$n=3$	$n=7$	P(2-tailed)*
Self-help general	9.21	8.00	7.83	7.93	<0.10
Self-help dressing	1.71	1.75	1.33	1.57	ns
Self-help eating	4.21	5.00	3.67	4.43	ns
Communication	3.00	3.12	3.00	3.07	ns
Socialization	3.07	3.00	2.33	2.71	<0.20
Locomotion	2.64	1.88	1.17	1.57	<0.05
Occupation	4.64	2.88	3.50	3.14	<0.02

* Based on t-ratio comparisons.

Appendix IV-8

Coefficients of correlation between the Mental and Motor Scale Quotients by age for 5-HTP treated and placebo groups as compared with Bayley's (1969) normal sample

Age (mth)	Subject group						
	Down's syndrome						Normal*
	5-HTP treated			Placebo			
	n	r_s	p	n	r_s	p	r
3	8	0.669	<0.10	8	0.926	<0.01	0.61
6	7	0.926	<0.01	8	0.803	<0.02	0.75
12	10	0.699	<0.05	8	0.890	<0.01	0.23
24	7	0.267	ns	7	0.449	ns	0.28
36	9	0.787	<0.02	9	0.302	ns	0.18 (30 mth)

* Bayley (1969).

Appendix IV-9

One year

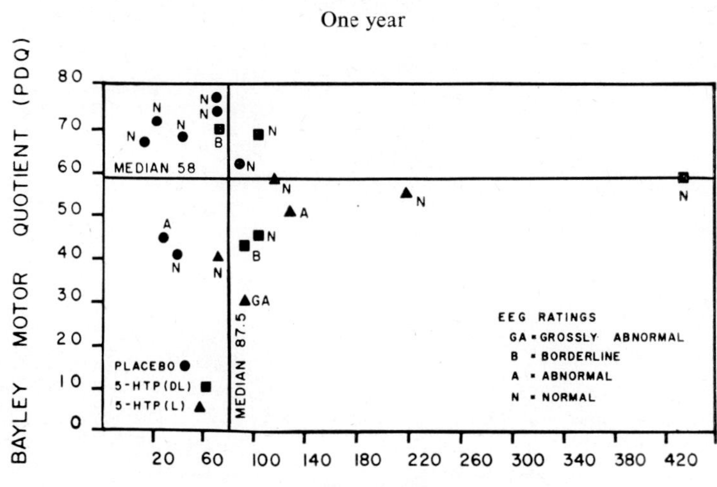

Two years

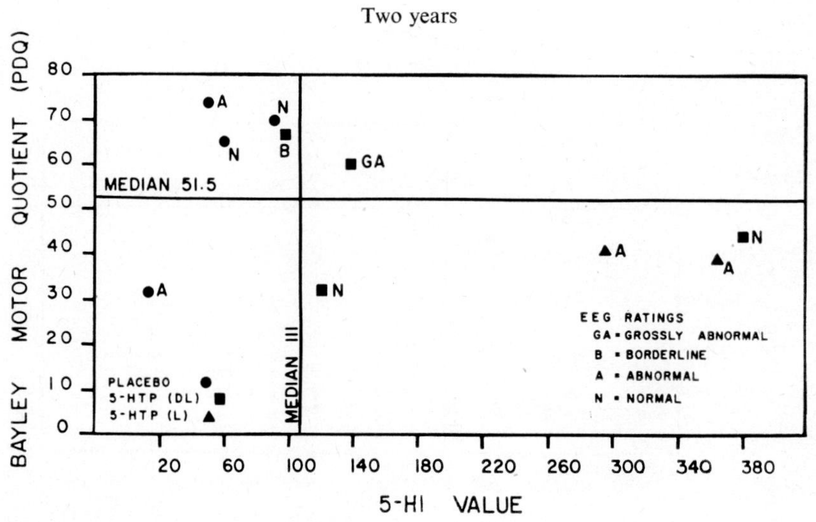

Three years

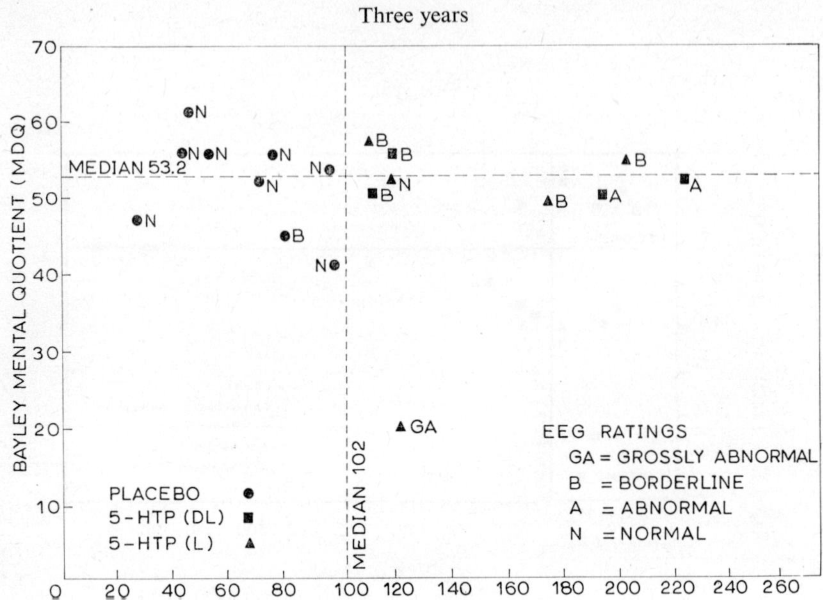

Three years

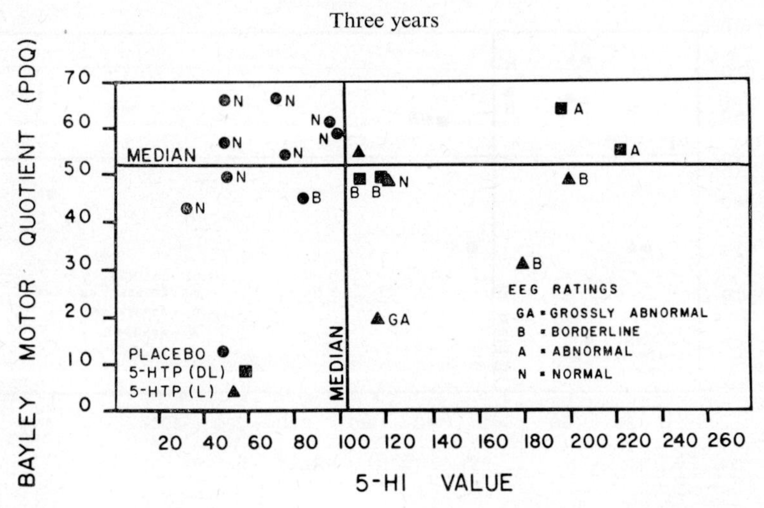

Appendix IV-10

Mean Bayley Mental and Motor Development Quotients for untreated, placebo and 5-HTP treated Down's syndrome infant groups at 3, 12 and 36 months of age

	Untreated		Placebo		5-HTP treated	
	MDQ	PDQ	MDQ	PDQ	MDQ	PDQ
3 mth	71.2	75.0	106.9	115.5	87.9	78.2
	$n=4$		$n=8$		$n=8$	
12 mth	69.2	65.2	69.8	62.8	62.3	50.6
	$n=6$		$n=8$		$n=10$	
36 mth	51.1	47.8	52.1	54.8	49.3	46.2
	$n=9$		$n=9$		$n=9$	

APPENDIX V

Appendix V-1

Personality inventory

Name of child.................................... Sex................ Race................ Age................
Name of examiner.................................... Date of observation....................................
State of the child (health, sleepiness, hunger, upsetting events, etc.)....................................
....
Number of people at home during observation....................................
Clinical remarks....................................

Scoring

(A) *Children*	*Range*	*Adj. score*
(1) Motor development and sturdiness	(7–35)	....
(2) Social relatedness	(15–75)	....
(3) Coping and mastery	(13–65)	....
(4) Emotional integration	(10–50)	....
Total score	(45–225)	....

(B) *Parents*	*Range*	*Adj. score*
(1) Acceptance of child	(1–5)	....
(2) Maternal responsiveness	(1–5)	....
(3) Maternal stimulation	(1–5)	....
(4) Maternal encouragement of independence	(1–5)	....
Total score	(4–20)	....

Appendix V-2

Children's rating

(1) *Motor development and sturdiness*

Item	Low	Scale	High	Level of confidence			Adj. score
Large muscle coordination	Poor	1–2–3–4–5	Good	1	2	3	
Small muscle coordination	Poor	1–2–3–4–5	Good	1	2	3	
Motor integration and stability	Poor	1–2–3–4–5	Good	1	2	3	
Energy level	Low, limp, passive	1–2–3–4–5	High	1	2	3	
Build	Fragile, thin, small	1–2–3–4–5	Age appropriate	1	2	3	
Muscle tone	Poor	1–2–3–4–5	Good	1	2	3	
Fatigability	Easily fatigued	1–2–3–4–5	Not easily fatigued	1	2	3	

(2) *Social relatedness*

Item	Low	Scale	High	Level of confidence			Adj. score
Visual alertness	Poor	1–2–3–4–5	Great	1	2	3	
Auditory alertness	Poor	1–2–3–4–5	Great	1	2	3	
Imitates vocalization	Doesn't imitate	1–2–3–4–5	Frequently	1	2	3	
Vocal response	Infrequent	1–2–3–4–5	Frequent	1	2	3	
Smiling	Infrequent	1–2–3–4–5	Frequent	1	2	3	
Efforts to imitate	Infrequent	1–2–3–4–5	Frequent	1	2	3	
Mirrors affective state	Infrequently	1–2–3–4–5	Frequently	1	2	3	
Participates in social games	Infrequently	1–2–3–4–5	Frequently	1	2	3	
Differentiation of approval and disapproval	None	1–2–3–4–5	Usually	1	2	3	
Seeks mutuality and closeness	Withdraws	1–2–3–4–5	Frequently	1	2	3	
Flexible acceptance of inhibitions and restrictions	Not usually	1–2–3–4–5	Usually	1	2	3	
Demands attention	Insatiable or extremely passive	1–2–3–4–5	Frequently	1	2	3	
Appearance	Repulsive	1–2–3–4–5	Appealing	1	2	3	
Initiates social interaction with family members	Infrequently	1–2–3–4–5	Frequently	1	2	3	
Initiates social interaction with observer	Infrequently	1–2–3–4–5	Frequently	1	2	3	

(3) *Coping and mastery*

				Level of confidence			*Adj. score*
Cooperation	Rejects or accepts non-selectively	1–2–3–4–5	Selective cooperation	1	2	3	
Assertion	Passive	1–2–3–4–5	Usually assertive	1	2	3	
Striving to improve skills and learn new ones	Very little	1–2–3–4–5	Frequently	1	2	3	
Persistence	Gives up easily or drives self to exhaustion	1–2–3–4–5	Good	1	2	3	
Functional use of objects	Poor	1–2–3–4–5	Good	1	2	3	
Independent play	Seldom or never	1–2–3–4–5	Frequent	1	2	3	
Attempts independent self care	Seldom	1–2–3–4–5	Frequently	1	2	3	
Performs for others	Seldom	1–2–3–4–5	Frequently	1	2	3	
Range of exploration	Narrow, constricted	1–2–3–4–5	Wide	1	2	3	
Goal-oriented	Not usually, aimless	1–2–3–4–5	Usually	1	2	3	
Alternates effort and rest	Seldom in constant motion or little activity	1–2–3–4–5	Usually	1	2	3	
Evidence of attachment to objects	Little	1–2–3–4–5	Good	1	2	3	
Ability to accept substitutes	Poor	1–2–3–4–5	Good	1	2	3	

(4) *Emotional integration (maturity)*

				Level of confidence			*Adj. score*
Emotional stability	Poor (tending to disorganize), volatile, irritable	1–2–3–4–5	Good, not easily upset	1	2	3	
Balance of developmental zones	Poor	1–2–3–4–5	Good	1	2	3	
Ability to accept change	Poor	1–2–3–4–5	Good	1	2	3	
Prevailing mood	Irritable, sad	1–2–3–4–5	Contented, happy	1	2	3	
Expressiveness	Constricted effect, monotone	1–2–3–4–5	Wide range of effect	1	2	3	
Frustration tolerance	Poor	1–2–3–4–5	Good	1	2	3	
Asks for help appropriately	Seldom or never	1–2–3–4–5	Usually	1	2	3	
Ability to be comforted by adult	Poor, inconsolable	1–2–3–4–5	Good	1	2	3	
Active avoidance of stressful situations	Seldom or never	1–2–3–4–5	Usually	1	2	3	
Ability to comfort self	Poor	1–2–3–4–5	Good	1	2	3	

Appendix V-3

Parent rating

	Levels of confidence			Adj. score
1. *Acceptance of child*				
5. Positive feelings about child most of the time	1	2	3	
4. Predominantly positive feelings but some interference from ambivalence	1	2	3	
3. Ambivalence about child is prominent	1	2	3	
2. Much criticism of child, though there is some warm feeling	1	2	3	
1. Clearly rejects	1	2	3	
2. *Maternal responsiveness (child initiates)*				
5. Mother responds quickly to child's verbal and non-verbal cues	1	2	3	
4. Mother responds most of the time	1	2	3	
3. Mother aware of child's needs but at times distracted with chores	1	2	3	
2. Mother responds only occasionally to the child and without much enthusiasm	1	2	3	
1. Mother not attuned to child, doesn't respond or understand his signals	1	2	3	
3. *Maternal interaction (areas as play, learning skills, social, etc.)*				
5. Fairly continuous stimulation of child	1	2	3	
4. Frequent interaction only by chores, etc.	1	2	3	
3. Interaction by timetable—mother has scheduled time with child	1	2	3	
2. Infrequent interaction—routine care—child left to own devices	1	2	3	
1. Interaction minimal—mother avoids interaction outside of routine care	1	2	3	
4. *Maternal encouragement of independence*				
5. Does not assist child unless absolutely necessary	1	2	3	
4. Occasionally assists to get child started on an activity	1	2	3	
3. Offers assistance immediately, when child asks for it	1	2	3	
2. Offers assistance, even though child doesn't ask for it	1	2	3	
1. Does everything for the child	1	2	3	

APPENDIX VI

Appendix VI-1

EEG sleep stages during sensory stimulation (% total sleep time)

Sleep stages	Placebo					D,L-5-HTP					L-5-HTP				
Age	0–9 days	6 mth	1 yr	2 yrs	3 yrs	0–9 days	6 mth	1 yr	2 yrs	3 yrs	0–9 days	6 mth	1 yr	2 yrs	3 yrs
N	5	3	7	5	9	5	3	5	3	4	4	2	4	2	4
SREM	20	23	26	0	4	32	14	8	26	1	32	8	7	0	2
1+indeterm.	28	7	32	34	39	9	14	33	14	25	25	2	16	50	45
SWS+TA or 2	52	59	40	43	39	59	56	46	31	33	43	73	45	0	19
3 and 4	0	11	2	24	18	0	16	13	29	41	0	17	27	50	34

Sleep stages:

SREM=Stage REM.

1+indeterm.=Stage 1+indeterminate sleep.

SWS+TA or 2=slow wave sleep plus tracé alternant or Stage 2. The classification SWS+TA was used for patients from 0–9 days only.

3 and 4=Stages 3 and 4.

The records of patients B-IS-19 and B-3 which showed hypsarrhythmia are omitted.

Appendix VI-2

Ratings of EEG background activity

Groups	Sex	Age groups					Totals 6 mth–3 yrs	
		Pre-treatment 0–10 days	6 mth	1 yr	2 yrs	3 yrs	#	%
Placebo								
B-2	M	N	N	N	A	N		
B-4	M	—	N	N	—	N		
B-5	F	N	—	N	N	N		
B-6	F	N	—	N	—	B		
B-12	M	N	N	N	—	N		
B-13	F	N	—	N	N	N		
B-14	M	—	—	—	—	N		
B-15	F	N	—	N	N	N		
B-16	F	—	—	N	A	N		
No. of records		6	3	8	5	9	25	100.0
Abnormal		0	0	0	2	0	2	8.0
Borderline		0	0	0	0	1	1	4.0
Normal		6	3	8	3	8	22	88.0
D,L-*5-HTP*								
B-1	F	N	N	N	N	B		
B-3	F	N	N	B	GA	A		
B-7	F	N	—	B	B	A		
B-8	F	N	N	N	N	B		
B-11	F	N	—	N	—	—		
L-*5-HTP*								
B-9	M	N	N	N	—	B		
B-10	F	N	N	N	A	B		
B-17	F	N	—	A	A	B		
B-18	M	—	—	N	—	N		
B-IS-19	F	N	GA	GA	GA	GA		
No. of records		9	6	10	7	9	32	100.0
Abnormal		0	1	2	4	3	10	31.2
Borderline		0	0	2	1	5	8	25.0
Normal		9	5	6	2	1	14	43.8

Ratings: N=normal EEG; B=borderline abnormal; A=abnormal; GA=grossly abnormal (hypsarrhythmia).

Appendix VI-3

Amplitude means and standard deviations of AER components obtained from 5-HTP treated and untreated Down's syndrome infants

Groups			5-HI ng/ml		Amplitudes (μV)											
					P_1N_1			N_1P_2			P_2N_2			N_2P_3		
Age	Treatment	*n*	$\bar{x}$	SD	*n*	$\bar{x}$	SD	*n*	$\bar{x}$	SD	*n*	$\bar{x}$	SD	*n*	$\bar{x}$	SD
10 days	Pre-placebo	4	32.2	8.0	4	10.5	7.5	4	23.4	12.8	4	29.8	10.1	3	8.8	3.6
	Pre-5-HTP	6	29.3	10.4	5	10.9	7.8	6	20.8	11.2	6	23.3	11.1	6	7.7	4.1
6 mth	Placebo	3	87.0	11.5*	3	4.6	3.8	3	20.0	21.5*	3	41.8	18.8*	3	24.2	10.4
	5-HTP	4	151.8	86.2	3	6.4	4.1	3	18.7	4.0	4	33.7	4.3	4	23.0	9.6
1 yr	Placebo	6	57.3	22.0*	6	6.5	1.4*	6	10.9*	3.9*	6	28.0	7.7	6	27.3	8.4
	5-HTP	7	162.7	127.7	7	5.4	4.5	7	17.1	8.9	7	37.9	8.9	6	31.1	11.8
2 yrs	Placebo	4	68.2*	17.8	3	4.0	2.1	3	9.5*	3.3	4	35.6*	3.9*	4	28.9	6.1
	5-HTP	3	121.0	20.5	3	7.3	5.2	3	25.3	4.9	3	64.7	13.7	3	42.1	9.8
3 yrs	Placebo	9	66.8*	24.2*	9	5.2	4.0	9	14.7	9.1	9	38.0	12.7	9	35.1	9.4
	5-HTP	8	159.0	46.5	6	5.0	5.5	6	11.3	7.9	8	44.8	15.3	8	41.4	10.8

n Denotes number of records with particular component.
* Measures where significant differences occurred between placebo and 5-HTP treated groups.

Appendix VI-4

Latency means and standard deviations of AER components obtained from 5-HTP treated and untreated Down's syndrome infants

Groups			Latencies of components (msec)														
Age	Treatment	*N*	P_1			N_1			P_2			N_2			P_3		
			n	$\bar{x}$	SD	*n*	$\bar{x}$	SD	*n*	$\bar{x}$	SD	*n*	$\bar{x}$	SD	*n*	$\bar{x}$	SD
0–10 days	Pre-placebo	4	4	74.6	16.3	4	145.3	10.6	4	256.8	43.6	4	507.1	11.1*	3	692.9	71.9
	Pre-5-HTP	6	5	90.6	15.4	6	152.2	18.7	6	269.4	26.9	6	500.5	43.2	6	714.6	82.6
6 mth	Placebo	3	3	74.8	11.6	3	108.0	13.2	3	175.7	28.2	3	374.7	106.5	3	707.5	82.0
	5-HTP	4	3	65.5	4.7	3	98.5	16.4	4	187.8	12.9	4	380.4	76.5	4	619.2	116.8
1 yr	Placebo	6	6	60.3	5.0*	6	93.8	7.9	6	178.7	17.6	6	350.6	17.0*	6	688.4	119.0*
	5-HTP	7	7	60.1	15.3	7	89.9	15.5	7	172.5	9.2	7	353.8	47.9	6	580.8	39.1
2 yrs	Placebo	4	3	58.2	10.8	3	81.8	7.4	4	165.1	1.5*	4	323.6	23.3	4	548.6	110.5
	5-HTP	3	3	63.8	37.1	3	93.6	31.4	3	163.1	23.4	3	298.2	41.2	3	517.0	155.5
3 yrs	Placebo	9	9	52.3	5.2	9	82.2	12.9	9	148.0*	18.1	9	328.0	57.8	9	736.7	92.5*
	5-HTP	8	6	51.0	8.7	6	75.7	6.8	8	165.2	13.6	8	320.6	51.3	8	664.0	174.0

n Denotes number of records with particular component.
* Measures where significant differences occurred between placebo and 5-HTP treated groups.

Appendix VI-5

Mean VER amplitudes (μV) of specific deflections

Groups			(−) (+) A–B			(+) (−) B–C			(−) (+) C–D			(+) (−) D–E			(−) (+) E–F			(+) (−) F–G			(+) (−) D–G		
Age	Treatment	N	n	$\bar{x}$	s	n	$\bar{x}$	s	n	$\bar{x}$	s	n	$\bar{x}$	s	n	$\bar{x}$	s	n	$\bar{x}$	s	n	$\bar{x}$	s
0–10 days	Pre-placebo	5	—	—	—	4	1.6	0.7	3	2.5	1.9	1	0.3	—	1	3.1	—	3	14.4	5.5	2	8.0	0.5
1 yr	Placebo	3	1	1.3	—	2	11.0	0.6	2	7.5	1.8	2	9.4	1.8	2	1.8	0.2	2	16.0	4.8	1	19.4	—
	5-HTP	6	4	4.0	4.0	6	10.5	11.8	6	11.9	10.0	1	3.8	—	1	1.6	—	1	19.8	—	5	41.1	10.4
3 yrs	Placebo	6	3	4.1	3.6	2	3.0	0.2	1	1.7	—	1	3.1	—	1	3.2	—	3	23.9	13.3	1	39.8	—
	5-HTP	5	1	0.8	—	3	6.0	4.5*	3	16.1	2.6	2	14.4	1.9	3	2.8	1.7	3	33.3	27.7	2	39.2	0.4

n Denotes numbers of recordings showing specific peaks.
* Measures where significant differences occurred between placebo and 5-HTP treated groups.

Appendix VI-6

VER peak latency tabulations of 3-year-old placebo and 5-HTP treated Down's syndrome patients

Subjects	Peak latencies (msec)						
	(−) A	(+) B	(−) C	(+) D	(−) E	(+) F	(−) G
Placebo							
B-13	24.5	67.4*	—	—	—	—	205.8
B-15	28.2	—	—	98.2*	—	—	221.0
B-6	—	50.2	82.1	—	—	153.1*	284.2
B-16	—	—	60.0	83.3	113.9	165.4*	232.8
B-2	30.4	55.7	76.0	—	—	167.1*	—
B-5	42.8	54.6	—	—	115.2	137.8*	207.9
5-HTP							
B-1	40.6	55.8	74.9	114.2*	152.3	167.5	223.3
B-17	—	53.4	62.3	110.7*	—	—	209.9
B-7	—	48.9	66.8	90.7*	115.7	134.8	219.5
B-3	44.4	—	—	114.2*	—	—	224.6
B-10	—	—	—	—	119.0	146.4*	219.0

* Latency of 'most prominent' positive deflection (P′).

APPENDIX VII

Appendix VII-1

Acute side effects of 5-hydroxytryptophan in patients in this study

	Patient		5-HTP		5-HI (ng/ml)	Percentage of total
	Code	Age	Type	Dosage (mg/kg)		
Gastrointestinal symptoms						
Diarrhea	27 patients	All ages	Both	All dosages	88–306	43
Gastric pain after dose	A-4	4 yrs	D,L		125	1.6
CNS Symptoms						
Overactivity or hyperactivity	D-12	6 wk	D,L		111	
	D-20	7 wk	D,L		92	
	A-7	2 mth	L		120	
	D-8	2 mth	D,L			
	B-19	2 mth	L		92	
	B-3	2 mth	D,L		132	
	B-7	2½ mth	D,L		119	
	D-13*	3½ mth	L		74	
	A-10	7½ mth	D,L		100	
	A-5	13 mth	D,L			
	A-13	20 mth	D,L			
	B-10	2½ yrs	L		144	
	A-12	2½ yrs	D,L		466	21
Opisthotonos	B-19	6 wk	L		97	
	B-17	2 mth	L		184	
	D-9*	2 mth	D,L		84	
	B-3	3 mth	L		118	
	A-4	3½ mth	D,L		150	
	B-1	4½ mth	D,L		86	10
Hyperacousis	B-1	6 wk	D,L	3.1	133	
	D-1	14 mth	D,L		112	
	A-13	4 yrs	D,L		198	
	A-5	5 yrs	D,L		78**	6
Vascular symptoms						
Hypertension	A-7	6 days	L	0.5	30	
	A-8	14 days	L	1.0	20	
	A-5	30 days	D,L	0.2	175	
	D-25		D,L			
	D-1		D,L			8
Acrocyanosis	A-12	3½ mth	D,L	98.7	318	1.6

* Also on B6. ** Missed two doses.

Appendix VII-2

Chronic side effects of 5-hydroxytryptophan in patients in this study

	Patient		5-HTP		5-HI ng/ml	Per-centage of total
	Code	Age	Type	Dosage (mg/kg)		
Clinical convulsive phenomena						
(a) EEG hypsarrhythmic	B-IS-19	5 mth	L	2.7	108	
	D-7	7 mth	D,L	1.5	125	
	C-32	11 mth	D,L	7.0	140	
	D-15	12 mth	D,L	6.6	264	
(b) EEG normal	D-4	12 mth	D,L	1.5	144	
	D-7	11 mth	D,L	6.8	258	
	D-13	9 mth	L	2.3	76	
	A-7	23 mth	L	2.0	122	
	A-8	22 mth	L	2.5	190	
	B-10	4 days	L	0.2	24	
	D-14	6 wk	D,L		125	17%
? Glaucoma	A-3	Diagnosed 3½ yrs	D,L		276	
	D-22	4 mth	L	Off 5-HTP 2½ mth	80	
? Ectodermal lesion						
(a) Alopecia areata	D-8	2 yrs	D,L		216	
(b) Granulomatous dermatitis	A-5	2 yrs, 8 mth	D,L		335	

Appendix VII-3

Patients in this study with glaucoma

PATIENT A-3
Diagnosis Bilateral glaucoma.
Age First seen by an ophthalmologist at 3½ years of age.
D,L-*5-HTP* 4 days until 4 years of age.
Description As a result of the long-standing high intraocular pressure the eye had stretched to the point that myopia was very prominent and also cupping of the optic nerves was considerable. In addition to the cupping of the optic nerves the vessels were displaced nasally. Surrounding the optic nerves was a zone of atrophy but this was probably related to the myopia.
Treatment September, 1969—goniotomy, O. D.
November, 1969—goniotomy, O. S.
(5-HTP completely tapered off May, 1970)
June, 1971—goniotomy, O. D.

PATIENT D-22
Diagnosis Bilateral congenital glaucoma.
Age Diagnosed at 4 months of age.
L-*5-HTP* 2 days to 5 weeks of age.
Description Examination under anesthesia at 4 months of age shows (1) evidence of ocular

changes secondary to infantile glaucoma which consists of enlarged cornea and several breaks in the posterior layer of the cornea from stretching, (2) abnormal corneal epithelium with extensive hydropic changes bilaterally which are not characteristic of glaucoma, (3) normal intraocular tension.
Treatment Goniotomy not required.

Appendix VII-4

Patients in this study with ectodermal lesions

PATIENT D-9

Surgical pathology report of scalp biopsy.
Anatomic source of specimen Scalp.

Gross examination Specimen labeled 'scalp biopsy' and consists of a 0.9 × 0.3 cm ellipse of skin. There is a small indentation on the surface; otherwise, the specimen is unremarkable. Section submitted *in toto.*

Microscopic examination Sections demonstrate skin showing an area of diminution of hair follicles and sparse mononuclear cell infiltration surrounding the residual follicles.

Diagnosis Skin of scalp, consistent with alopecia areata.

PATIENT A-5

Surgical pathology report of skin of face biopsy.
Anatomic source of specimen Skin of face.

Gross examination The tissue is a tiny spindle-formed wedge measuring 0.45 × 0.2 cm, tapering to a depth of 0.2 cm. The epidermal surface is slightly rough and dull, while the underlying chorium has a moist, translucent, slightly bulging appearance.

Microscopic examination Sections of skin show a granulomatous infiltrate to become slightly less dense in the deeper portion. It thins the overlying epidermis, weaves between the thick collagen bundles of the dermis, and encircles the few adnexal structures present. The infiltrate is generally band-like and lacks a nodular pattern. No well formed giant cells are seen, the reaction consisting of mixed epithelioid cells, histiocytes and lymphocytes. No necrosis is seen within the granulomatous lesions and collagen necrosis is absent. No organisms or refractile material are seen. The sections of patient A-5 show indeed a granulomatous infiltrate which consists entirely of mononuclear cells of lymphocytic and histiocytic type. The sections are certainly not typical of either tuberculosis or sarcoidosis, but the infiltrate is too massive for a lichen scrofulosorum. A granulomatous drug eruption is a possibility but is difficult to confirm.

Diagnosis Skin of face—granulomatous dermatitis, non-specific.

APPENDIX VIII

Appendix VIII-1

Patient B-2

Placebo followed by pyridoxine studies

	Age (mth)	5-HI	Rx	5-HIAA (μg/kg)	HVA (μg/kg)	VMA (μg/kg)	Noreri (μg/kg)	Epi (μg/kg)
Age 1–2 yrs	14	37	Placebo	32.1	—	65	0.730	0.170
	16	66	Placebo					
	18	63	Placebo					
	21	51	Placebo					
Arithmetic mean		54	Placebo					
Age 2–3 yrs	25	53	Placebo	77.7	74	115	0.708	0.393
	28	65	Placebo					
	30	60	Placebo					
	34	65	Placebo					
	36	37	Placebo	14.31	34	62	—	—
Arithmetic mean		56	Placebo					
			mg/kg dose					
Age 3–4 yrs	38	74	B6— 5.0					
	40	50	B6— 5.0					
	41	101	B6— 6.5					
	44	86	B6— 8.0	11.98	156	101	—	—
	45	84	B6—10.0					
	46	88	B6—11.0	34.29	119	60	0.399	0.097
Arithmetic mean		82	Vitamin B6					

Appendix VIII-2

Patients on B6 prior to clinic visit compared to age-matched controls

First visit 5-HI levels in trisomy 21 patients

Age drawn (days)	5-HI (ng/ml)	
	Patient on B6	Age-matched control
3	25	7
3	37	25
4	24	31 (average of 5 pts.)
6	50	28 (average of 9 pts.)
6	51	28 (average of 9 pts.)
7	56	42 (average of 3 pts.)
7	75	42 (average of 3 pts.)
9	40	46 } only one control available
9	87	46 }
12	58	50 } only one control available
12	65	50 }
16	64	50 }
21	117	88
24	112	30
26	72	41
27	58	76
29	68	55
Arithmetic mean	62	40

APPENDIX IX

Appendix IX-1

Factors affecting serotonin in children

A. DISEASE ENTITIES (Primary CNS disease excluded, see Chapter 9)

1. *Gastrointestinal disease*
   a. Carcinoid disease
   b. Malabsorption syndrome
   c. Acute gastroenteritis
   d. Intestinal obstruction
   e. Congenital anomalies of gastrointestinal tract
   f. ?Functional gastrointestinal hyperserotoninemia syndrome

2. *Vascular disease*
   a. Collagen–vascular disease—active rheumatoid arthritis; Raynaud's syndrome
   b. Migraine
   c. (Following) myocardial infarction
   d. ?Myeloproliferative disease

3. *Thyroid*
   a. Hypothyroid (infants)
   b. Hyperthyroid

(Erspamer, 1966; Warner, 1967; Page, 1968; Coleman, 1970).

In some of these syndromes, the data are based on 5-HIAA excretion in urine rather than 5-HT levels in blood. There are scattered case reports in toxemia of pregnancy, shock, asthma, hepatic coma and cirrhosis of the liver. In many of these diseases, however, these 5-HT abnormalities are not consistently seen in all patients, suggesting it may not be a specific effect of the particular diseased organ system. Also, even in the carcinoid syndrome, the disease entity with the most abnormal 5-HT values, the levels in the blood do not correlate well with the severity of clinical symptoms (Levine and Sjoerdsma, 1963). The mechanism resulting in 5-HT abnormality appears to differ in each disease but, to date, in no disease entity has an error in 5-HT metabolism been definitely established as the primary pathogenic process. One reason for the large number of disease entities with recorded 5-HT abnormalities may be the fact that 5-HT shares enzymes, co-enzymes, transport and binding mechanisms with a number of other amino acid pathways, is altered by derangements in these other pathways and then is conveniently concentrated in the platelets where its fluctuations in levels can be readily extracted and measured.

B. BODY RHYTHMS

1. *Circadian rhythm* (Quay, 1968; Krieger and Rizzo, 1969). An 18% circadian variation in serum 5-HT levels in Down's syndrome and other retarded patients has been demonstrated (Halberg *et al.*, 1967). (In order to account for this factor, total 5-HI levels drawn on each individual patient in this monograph were drawn at the same hour of the morning throughout the years of the study.)
2. *Menstrual rhythms* (Ritvo *et al.*, 1971; Vogel *et al.*, 1970).

C. Diet

Fruits (including tomatoes and eggplant) and nuts are the major dietary sources of exogenous 5-HT. We had an opportunity to study the effect of diet on whole blood 5-HI levels in a psychotic 8-year-old whose food compulsion was focused that season on bananas, the food containing the highest proportion of 5-HT per milligram. Prior to the banana loading test, we were unable to detect any abnormality in the patient's endogenous 5-HI level in blood, 5-HT efflux from platelets, urinary 5-HIAA, HVA, VMA, MHPG, epinephrine and norepinephrine. This child consumed 1,301 g of peeled, whole ripe bananas, or 39.0 mg of 5-HT, during a one-hour lunch period (1.6 mg/kg of exogenous 5-HT.) A rise of 35 ng/mg of whole blood 5-HI occurred within an hour of ingestion; the effect was virtually gone by 4 hours. This is equivalent to approximately 3 ng/ml change of whole blood 5-HI per banana; factors of banana size and ripeness could increase this slightly.

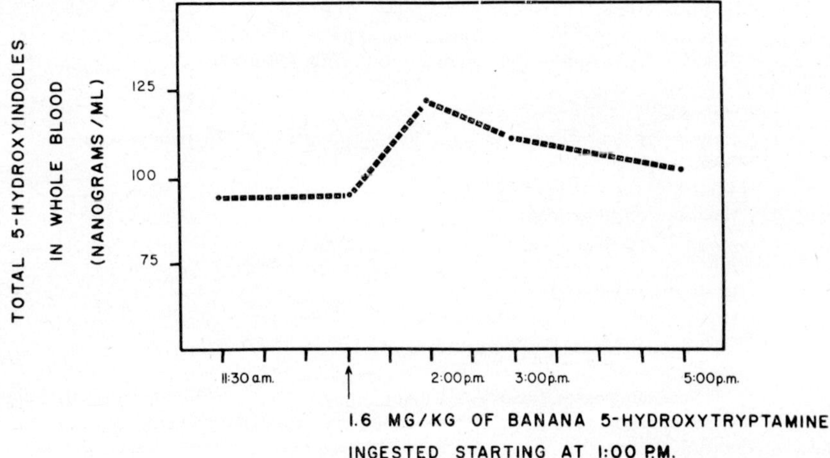

It is important to note that very little 5-HT consumed in food enters the central nervous system (Erspamer, 1966) because of the complex transport systems (formerly called the blood–brain barrier) that govern the entry of active amines into the brain; a small amount may enter by passive diffusion (Bulat and Supek, 1967). However, rat studies have shown that ingestion of food containing tryptophan or stimulating insulin may affect 5-HT levels in the CNS (Fernstrom and Wurtman, 1971).

D. Drugs

Pharmaceuticals used in children that affect serotonin

- amitriptyline
- amphetamines
- antihistamines
- barbiturates
- chloral hydrate
- diazepam
- diphenylhydantoin
- hydrocortisone, prednisone
- neomycin
- primidone
- purgatives

An extensive table of compounds, with experimental details of how 5-HT is affected, can be found in Appendices II and IV of Garattini and Valzelli (1965).

All the major psychotropic drugs affect 5-HT (phenothiazines, reserpine, tricyclic anti-depressants, monoamine oxidase inhibitors, amphetamines and lithium). Their complex mechanisms of action are reviewed in Shepherd *et al.* (1968) and Von Brücke *et al.* (1969) and a new mechanism of action has been added by Tagliamonte and colleagues (1971). The effect of L-Dopa on 5-HT is reviewed in Chapter 3.

Also, an investigator needs to be aware of the effects of drugs ingested by the patient upon laboratory procedures. In 1962, Pitkänen *et al.* described a patient operated on for a carcinoid syndrome because of a falsely elevated 5-HIAA level (method: Undenfriend *et al.*, 1955) caused by a metabolite of glyceryl guaiacolate, an ingredient in cough syrup (Pitkänen *et al.*, 1962). False positive 5-HIAA levels can also be seen with acetanilide, mephenesin and methocarbamol (Honet *et al.*, 1959; Pedersen *et al.*, 1970). Drugs which affect the total 5-hydroxyindole method in whole blood are reserpine and diazepam (increase fluorescence) and haloperidol (decrease fluorescence).

E. Stress

There is an extensive literature relating 5-HT changes to stress. The latest addition to the list of stressful situations is flying in satellite space-ships—it decreases 5-HT in mice and dogs (Parin *et al.*, 1965).

The preponderance of evidence suggests that raising the adrenocorticosteroids in *acute* stress may stimulate the rate-limiting, enzymatic step of serotonin metabolism (tryptophan hydroxylase) and cause increased turnover of serotonin in the brain; below normal turnover occurs when an animal is adrenalectomized (Azmitia *et al.*, 1970; Hanig *et al.*, 1970; Nistico and Preziosi, 1969). On the other hand, *chronic* stress in animals may eventually lower serotonin levels (Riege and Morimoto, 1970). In an experiment with hares, Miline *et al.* (1958) showed fear affected the 5-HT level in the brain. Initially 5-HT was raised (for 7 days): then it was reduced below baseline after 10 days.

Surgical procedures, anaphylactic shock, anoxia, hyperoxia, decompression, electro-shock, and X-radiation have all been demonstrated to affect 5-HT levels (Garattini and Valzelli, 1965). At least in rabbits, fever is not accompanied by increased 5-HT turnover.

Environment *per se* may also be an important variable. The results in this monograph are based on patients living at home and studied mostly as out-patients. Children living in institutions have such radically different life experiences than children living at home that this undoubtedly is a factor in interpretation of some of the biochemical abnormalities reported from institutionalized studies. For example, we have recently examined whole blood 5-HI levels in rhesus monkeys raised with lack of mothering and noted lower levels in the younger animals (Coleman, 1971). Changes in the enzymes of catecholamine synthesis have also been noted after similar environmental stress (Axelrod and Mueller, 1970).

Even a temporary change in a child's environment, such as admission to a hospital (separation from home, parents, friends and subjection to uncomfortable procedures) can affect 5-HI levels. Almost all children we have studied have an initial drop of whole blood 5-HI when first admitted to a hospital compared to a previous out-patient value. This applies to patients with out-patient low, normal or high 5-HI baselines. The value remains low for 24 to 72 hours and then returns to the usual baseline. An exception to this pattern is seen in hyperactive children classified as 'functional' rather than organic; they may elevate rather than depress whole blood 5-HI levels upon hospital admission (Coleman, 1971). This variation in 5-HI levels at the time of hospitalization has led to us collecting these values in the out-patient clinic rather than during hospitalizations. The same phenomena have been seen in studies of free fatty acids in hospitalized psychotic children—the levels are affected by the fact of hospitalization, apart from physical illness (DeMyer *et al.*, 1971).

The most difficult factors of all to identify are purely emotional changes correlated with whole blood 5-HI variations. However, there appears to be no other explanation of sudden large swings in 5-HI levels in some patients. Patients B-15, described on the first page of this monograph appears to be one example of a drop in 5-HI, secondary to an emotional factor. Another case was seen in an 11-month-old trisomy patient on the pyridoxine regimen (Chapter 8). Although the dose of vitamin B6 was unchanged, the patient's 5-HI level had dropped 47 ng/ml (from 96 ng/ml to 49 ng/ml) after his sister was killed in an accident. This was the lowest value ever recorded on this child, including pre-pyridoxine values. It seemed that the parents' depression was the initiating event in the falling 5-HI level in this child.

In studies on adult human beings, a relationship between lowered 5-hydroxyindole metabolism and clinical symptoms of depression has been postulated (Bourney *et al.*, 1968; Lapin and Oxenkrug, 1969; Pare *et al.*, 1969; Van Praag *et al.*, 1970), although 5-HT metabolism appears unchanged in the same psychotically depressed patient during clinically asymptomatic or clinically depressed priods (Coppen, 1970; Shaw *et al.*, 1971).

Bunney and his colleagues (1969) have suggested there may be two types of clinical depression—a retarded, slowed-down patient with catecholamine deficit and an anxious, agitated patient with a serotonin deficit. 'Functional' hyperactive children may belong to the agitated depressive category.

Because of the intimate sensitivity between a mother and a baby, evaluation of a mother's emotional state occasionally may be necessary as part of the interpretation of 5-HI values in a young child.

SUMMARY

Clinical relevance, yet complete non-specificity, is the mark of the serotonin platelet model. Its value in the detection of retardation and clinical monitoring of the hyperactivity syndromes may be established. Yet, the interpretation of an abnormal level is complex in most patients, limiting its value clinically.

Appendix IX-2

Retarded patients with reported abnormal serotonin levels in blood

	5-HT level in blood	Method used for determination
1. *Two or more authors agree on the 5-HT abnormality*		
Down's syndrome	Low	Platelet Whole blood
Phenylketonuria	Low	Serum Platelet Whole blood
Histidinemia	Low	Platelet
Infantile spasm syndrome	High	Whole blood
2. *One author has two or more cases*		
Infant hypothyroidism	High	Whole blood
Maternal rubella	High	Serum
Kernicterus	High	Serum
deLange syndrome	Low	Whole blood
3. *Single case reports*		
Schilder's disease	High	Whole blood
Gargoylism	High	Serum
Sturge Weber (a retarded patient)	High	Serum
'Cerebral lipidosis'	High	Serum
Sudanophilic leukodystrophy	High	Whole blood
4. *Conflicting reports of single cases (two authors disagree)*		
Dysautonomia (retarded patient?)	High Normal	Serum Whole blood
Tuberous sclerosis	High Normal	Serum Whole blood

REFERENCES

AXELROD, J. and MUELLER, R. A. (1970) Changes in enzymes involved in the biosynthesis and metabolism of noradrenaline and adrenaline after psychosocial stimulation. *Nature*, **225**, 1059.

AZMITIA, E. C., ALGERI, S. and COSTA, E. (1970) *In vivo* conversion of ^{3}H-L-tryptophan into ^{3}H-serotonin in brain areas of adrenalectomized rats. *Science*, **169**, 201.

BOURNE, H. R., BUNNEY, W. E., COLBURN, R. W., DAVIS, J. M., DAVIS, J. N., SHAW, D. M. and COPPEN, A. J. (1968) Noradrenalin, 5-hydroxytryptamine and 5-hydroxyindole-acetic acid in hindbrains of suicidal patients. *Lancet*, **2**, 805.

BULAT, M. and SUPEK, Z. (1967) The penetration of 5-hydroxytryptamine through the blood–brain barrier. *J. Neurochem.* **14**, 265.

BUNNEY, W. E., JANOWSKY, D. S., GOODWIN, F. K., DAVIS, J. M., BRODIE, H. K. H., MURPHY, D. L. and CHASE, T. N. (1969) Effect of L-Dopa on depression. *Lancet*, **1**, 885.

COLEMAN, M. (1970) Serotonin levels in infant hypothyroidism. *Lancet*, **2**, 365.

COLEMAN, M. (1971) Serotonin concentrations in whole blood of hyperactive children. *J. Pediat.* **78**, 985.

COPPEN, A. (1970) Pituitary-adrenalin activity during psychosis and depression. *Progr. Brain Res.* **32**, 336.

DEMYER, M., SCHWIER, H., BRYSON, C., SOLOW, E. and ROESKE, N. (1971) Free fatty acid response to insulin and glucose stimulation in schizophrenic, autistic and emotionally disturbed children. *J. Aut. Child. Schizo.* **1**, 436.

ERSPAMER, V. (1966) 5-Hydroxytryptamine and related indolealkylamines, Springer-Verlag, New York.

FERNSTROM, J. D. and WURTMAN, R. J. (1971) Brain serotonin content: increase following ingestion of carbohydrate diet. *Science*, **174**, 1033.

GARATTINI, S. and VALZELLI, L. (1965) Serotonin, Elsevier, Amsterdam.

HALBERG, F., ANDERSON, J. A., ERTEL, R. and BERENDES, H. (1967) Circadian rhythm in serum 5-hydroxytryptamine of healthy men and male patients with mental retardation. *Int. J. Neuropsychiat.* **3**, 379.

HANIG, J. P., AIELLO, E. L. and SEIFTER, J. (1970) The effects of stress and intravenous 0.9% NaCl injection on concentrations of whole brain 5-hydroxytryptamine in the neonate chick. *J. Pharm. Pharmacol.* **22**, 317.

HONET, J. C., CASEY, T. V. and RUNYAN, J. W. (1959) False-positive urinary test for 5-hydroxyindoleacetic acid due to methocarbamol and mephenesin carbamate. *New Engl. J. Med.* **261**, 188.

KRIEGER, D. T. and RIZZO, F. (1969) Serotonin mediation of circadian periodicity of plasma 17-hydroxycorticosteroids. *Amer. J. Physiol.* **217**, 1703.

LAPIN, I. P. and OXENKRUG, G. F. (1969) Intensification of the central serotonergic processes as a possible determinant of the thymoleptic effect. *Lancet*, **1**, 132.

LEVINE, R. J. and SJOERDSMA, A. (1963) Pressor amines and the carcinoid flush. *Ann. Intern. Med.* **58**, 818.

MILINE, R., STERN, P. and HUKOVIC, S. (1958) Sur les variations stressogènes quantitatives de la sérotonine dans le cerveau. *Experientia*, **14**, 415.

NISTICO, G. and PREZIOSI, P. (1969) Brain and liver tryptophan pathways and adrenocortical activation during restraint stress. *Pharmac. Res. Comm.* **1**, 363.

PAGE, I. H. (1968) Serotonin, Year Book Medical Publishers, Inc., Chicago.

PARE, C. M. B., YEUNG, D. P. H., PRICE, K. and STACEY, R. S. (1969) 5-hydroxytryptamine, noradrenalin and dopamine in the brainstem, hypothalamus, and caudate nucleus of controls and of patients committing suicide by coal gas poisoning. *Lancet*, **2**, 133.

PARIN, V. V., ANTIPOV, V. V., RAUSHENBAKH, M. O., SAKSONOV, P. P., SHASHKOV, V. S. and CHERNOV, G. A. (1965) Changes in blood serotonin level in animals exposed to ionizing radiation and dynamic factors of space flight. *Izv. Akad. Nauk. SSSR* [*Biol.*], **30**, 3.

PEDERSEN, A. T., BATSAKIS, J. G., VANSELOW, N. A. and MCLEAN, J. A. (1970) False-positive tests for urinary 5-hydroxyindoleacetic acid. Error in laboratory determinations caused by glyceryl guaiacolate. *J. Amer. Med. Ass.* **211**, 1184.

PITKÄNEN, E., AIRAKSINEN, M. M., MUSTALA, O. O. and PALONEIMO, J. (1962) Observations of the specificity of the urinary 5-hydroxyindoleacetic acid determination with 1-nitroso-2-naphthol. *Scand. J. Clin. Lab. Invest.* **14**, 571.

QUAY, W. B. (1968) Differences in circadian rhythms in 5-hydroxytryptamine according to brain region. *Amer. J. Physiol.* **215**, 1448.

RIEGE, W. H. and MORIMOTO, H. (1970) Effects of chronic stress and differential environments upon brain weights and biogenic amine levels in rats. *J. Comp. Physiol. Psychol.* **71**, 396.

RITVO, E., YUWILER, A., GELLER, E., PLOTKIN, S., MASON, A. and SAEGER, K. (1971) Maturational changes in blood serotonin levels and platelet counts. *Biochem. Med.* **5**, 90.

SHAW, D. M., MACSWEENEY, D. A., WOOLCOCK, N. and BEVAN-JONES, A. B. (1971) Uptake and release of ^{14}C-5-hydroxytryptamine by platelets in affective illness. *J. Neurol. Neurosurg. Psychiat.* **34**, 224.

SHEPHERD, M., LADER, M. and RODNIGHT, R. (1968) Clinical psychopharmacology, Lea and Febiger, Philadelphia.

TAGLIAMONTE, A., TAGLIAMONTE, P., PEREZ-CRUET, J., STERN, S. and GESSA, G. L. (1971) Effect of psychotropic drugs on tryptophan concentration in the rat brain. *J. Pharmacol. Exp. Ther.* **177**, 475.

UDENFRIEND, S., TITUS, E. and WEISSBACH, H. (1955) The identification of 5-hydroxy-3-indoleacetic acid in normal urine and a method for its assay. *J. Biol. Chem.* **216**, 499.

VAN PRAAG, H. M., KORF, J. and PUITE, J. (1970) The influence of probenecid on the concentration of 5-hydroxyindoleacetic acid in the cerebrospinal fluid in depressive patients. *Nature* (*London*), **225**, 1259.

VOGEL, S. A., JANOWSKY, D. J. and DAVIS, S. M. (1970) Effect of estradiol on stimulus-induced release of ^{3}H-norepinephrine and ^{3}H-serotonin from rat brain slices. *Res. Comm. Chem. Path. Pharmacol.* **1** (**4**), 451.

VON BRÜCKE, F. Th., HORNYKIEWICZ, O. and SIGG, E. B. (1969) The pharmacology of psychotherapeutic drugs, Springer-Verlag, New York.

WARNER, R. R. P. (1967) Current status and implications of serotonin in clinical medicine. *Adv. Int. Med.* **13**, 241.

Subject Index

acrocyanosis 117, 120
ACTH 121, 122, 124, 143, 168
adenosine triphosphate (ATP) 19, 138, 156
aldehyde dehydrogenase 8
alkaline phosphatase 20
amniocentesis 2
alopecia areata 125, 126, 131
anemia 126
aromatic L-amino acid decarboxylase 8, 13, 17, 53, 139, 142
ataxia 29, 30, 49, 119
'attitudinal effect' 50, 59, 166
auditory information processing 64
autism 149, 156, 157

Bayley Scales of Mental and Motor Development 66, 67, 68, 74, 76, 77, 78, 79, 80, 81, 82, 166, 167, 168
  Mental Development Quotients 69, 70
  Psychomotor Development Quotients 71, 72, 73
Bayley Infant Behavior Profile 67, 74, 75, 76

carcinoid tumor 8, 154
catechol amines 9, 10, 36, 37, 53, 54, 55, 158, 168
Cattel Infant Scale 140
'cerebral palsy' 150, 155–156
chloride ions 29
constipation 158

deoxypyridoxine 139
depression 5, 21, 157, 158, 160, 161
desipramine 145
diarrhea 117, 118, 127, 168
diphenylhydantoin 159
dopamine 9
dopamine-beta-hydroxylase 55
double blind study of 5-HTP 43–114
double chromosomal error 11, 12, 15, 16, 142, 166
Down's syndrome
  age of walking 30, 31, 38, 59
  buccal-lingual dyskinesia 33, 34, 38, 51, 52, 59, 158, 167
  buccal-lingual hypotonia 33, 34, 38, 51, 52, 158, 167
  cardiac disease 32, 33, 38, 51, 59, 129, 167
  cerebellar function 29, 30
  cranial circumference 38, 55, 59, 76, 77, 167
  incidence 2
  height 38, 55, 159, 167
  intelligence 1, 2, 6, 61–83, 92, 140
  muscle tone 6, 27, 28, 29, 30, 38, 43, 46, 47, 48, 49, 50, 59

Down's syndrome—contd.
pathology 3
personality patterns 6, 87–93
prognosis 2
strabismus 32, 38, 50, 51, 59, 167
temperature patterns 6, 36, 52, 53, 59, 167
weight 38, 55, 59, 167
dysautonomia 155
dystonia 127

ectodermal lesions 117, 125
electroencephalogram (EEG) 6, 32, 48, 79, 82, 92, 95–114, 129, 130, 143, 166, 167, 168
hypsarrhythmia 98, 99, 114, 121, 122, 124, 129, 143, 152, 166
sleep stages 96, 97, 99, 100–103, 161, 167
enterochromaffin cells of the gastrointestinal tract 6, 8, 17, 19
environmental stimulation 2, 62, 64, 87
epinephrine 9, 37, 53, 54, 168
evoked cortical potentials 95, 96, 103
auditory 65, 104–109, 166
visual 65, 109–113, 166

'false neurotransmitter' effect 138

GABA aminotransferase 143
GA decarboxylase 143
galactosemia 66, 150
Gesell Scale 63
glaucoma 117, 125, 130
glucose-6-phosphate dehydrogenase (G-6-PD) 20

head banging 56, 58, 59
histidinemia 149, 150, 152, 154, 160
home-reared *vs* institutionalization 61
homovanillic acid (HVA) 9, 53, 59, 168
hydrocephalus 157, 160
3-hydroxyanthranilic acid 153
3-hydroxykynurenine 153
5-hydroxyindoleacetic acid (5-HIAA) 8, 12, 17, 26, 55, 127, 136, 137, 138, 155, 156, 157, 159, 168
5-hydroxyindoles (5-HI) 5, 9, 10, 11, 12, 13, 14, 15, 16, 17, 18, 20, 21, 26, 35, 36, 44, 45, 46, 48, 55, 56, 57, 58, 77, 78, 87, 103, 106, 107, 108, 111, 112, 113, 114, 119, 124, 130, 136, 137, 138, 139, 141, 142, 143, 144, 145, 146, 154, 155, 161, 165, 168
5-hydroxytryptamine (5-HT)
efflux from platelets 20, 21
uptake into platelets 19, 21
location in body 6, 7, 8
metabolic pathway 7, 9
methods 8, 9, 12
platelet model system 37, 150, 151, 160
5-hydroxytryptophan (5-HTP)
double blind study of 43–114
side effects 117–132
also 8, 12, 17, 25, 26, 27, 28, 29, 30, 31, 32, 33, 35, 36, 37, 38, 139, 141, 143, 165, 166, 167, 168
hyperacousis 117, 119, 128, 168
hyperactivity 117, 118, 128, 143, 149, 157, 168
hyperphenylalaninemia 150
hypertension 117, 120, 128, 168
hyperthyroidism 154
hyperuricemia 127
hypothyroidism 149, 150, 154, 158
hypotonia 28, 30, 34, 44, 48, 63, 65, 66, 124, 149, 159, 160, 161, 166

idiot savant syndrome 131, 157

imipramine 144–145
infantile spasm syndrome 98, 118, 121–124, 128–130, 132, 143, 149, 150, 152, 153, 154, 166, 168
infant stimulation 2
iproniazid 135, 136

kernicterus 149, 155
kynurenine 17

Landau posture 27, 28, 49, 158, 159
L-DOPA 25, 27, 28, 29, 30, 35, 36, 37, 43, 53, 127, 131

melatonin 8
methoxy-4-hydroxyphenylethyleneglycol (MHPG) 9, 53, 59, 168
methyl-lysergic acid-butanolamide 154
monoamine oxidase 8, 18, 131, 136
monoamine oxidase inhibitors 135–138, 145, 146, 168
Moro reflex 27
mosaics 1, 2, 10, 12, 13, 14, 21, 27, 28, 44, 63, 165

N-acetylserotonin 8, 138
nialamide 135
norepinephrine 9, 53, 54, 55, 59, 168
normetanephrine 136

opisthotonus 117, 119, 168

Parkinson's disease 25, 35, 36, 43, 127
p-chlorophenylalanine (*p*CPA) 29, 127
pelvic radiation in parents 44, 159
D,L-penicillamine 142
perceptuocognitive function 64, 66
personality inventory 87–93, 167
phenotypical mongols 10, 16, 21, 166
phenylalanine 7, 37, 150
phenylketonuria (PKU) 66, 149, 150, 151–152, 154, 160
phosphofructokinase 20
platelet model system for 5-HT 37, 150, 151, 160
platelets 9, 17, 18, 19, 55
potassium ions 29
probenecid 18, 143
pyridoxal kinase 144
pyridoxal-5-phosphate (PLP) 138, 140, 144, 153
4-pyridoxic acid (4-PA) 139
pyridoxine, *see* vitamin B_6

reserpine 160
retardation, non-specific 156
RNA 37
rubella, maternal 149, 155

seizures 98, 99, 114, 118, 121–124, 128–130, 143, 152–154, 159, 161, 166
sensorimotor experience 65, 66
serotonin, *see* 5-hydroxytryptamine
serotonin-*O*-glucuronide 138
serotonin syndromes of infancy 158–160, 168
Stanford Binet Testing 67, 68, 70, 71

taurine 140
thyroid 154
thyroxine 154
traction response 27
translocations 1, 2, 10, 12, 14, 21, 44, 63, 140, 143, 165
tranylcypromine 136–138, 146, 168
trisomy 21 1, 2, 10, 11, 12, 17, 20, 21, 26, 165
  double blind study of 5-HTP 43–114
tryptamine 136

tryptophan 7, 8, 17, 18, 25, 26, 129, 130, 139, 141, 142, 144, 145, 146, 152, 153, 168
tryptophan hydroxylase 7, 13, 17, 29
tuberous sclerosis 153, 155
p-tyramine 136
tyrosine 139

vanilmandelic acid (VMA) 9, 53, 59, 168
Vineland Social Quotient 63, 67, 73, 167
visual perception 64
visuomotor function 64
vitamin B_6 8, 13, 17, 18, 27, 83, 129, 138–144, 145, 146, 153, 168

xanthurenic acid 153